From Filing and Fitting to Flexible Manufacturing

From Filing and Fitting to Flexible Manufacturing

Roger E. Bohn

*Graduate School of International Relations
and Pacific Studies
University of California, San Diego
La Jolla, CA 92093-0519,
USA*

Rbohn@ucsd.edu

Ramchandran Jaikumar

(1944-1998)

the essence of knowledge
Boston – Delft

Foundations and Trends® in Technology, Information and Operations Management

Published, sold and distributed by:
now Publishers Inc.
PO Box 1024
Hanover, MA 02339
USA
Tel. +1 (781) 871 0245
www.nowpublishers.com
sales@nowpublishers.com

Outside North America:
now Publishers Inc.
PO Box 179
2600 AD Delft
The Netherlands
Tel. +31-6-51115274

A Cataloging-in-Publication record is available from the Library of Congress.

Printed on acid-free paper

ISBN: 1-933019-06-9; ISSNs: Paper version 1571-9545; Electronic version 1571-9553
© 2005 Now Publishers

Foundations and Trends® in Technology, Information and Operations Management

Volume 1 Issue 1-2, 2005

Editorial Board

Editorial Scope

Foundations and Trends® in Technology, Information and Operations Management will publish survey and tutorial articles in the following topics:

- B2B Commerce
- Business Process Engineering and Design
- Business Process Outsourcing
- Capacity Planning
- Competitive Operations
- Contracting in Supply Chains
- Electronic markets, auctions and exchanges
- Enterprise Management Systems
- Facility Location
- Information Chain Structure and Competition
- International Operations
- Marketing/Manufacturing Interfaces
- Multi-location inventory theory
- New Product & Service Design

- Queuing Networks
- Reverse Logistics
- Service Logistics and Product Support
- Supply Chain Management
- Technology Management and Strategy
- Technology, Information and Operations in:
  - Automotive Industries
  - Electronics Manufacturing
  - Financial Services
  - Health Care
  - Industrial Equipment
  - Media and Entertainment
  - Process Industries
  - Automotive Industries
  - Retailing
  - Telecommunications

Information for Librarians

Foundations and Trends® in Technology, Information and Operations Management, 2005, Volume 1, 4 issues. ISSN paper version 1571-9545 (USD 300 N. America; EUR 300 Outside N. America). ISSN online version 1571-9553 (USD 300 N. America; EUR 300 Outside N. America). Also available as a combined paper and online subscription (USD 340 N. America; EUR 340 Outside N. America).

Contents

FROM FILING AND FITTING TO FLEXIBLE
MANUFACTURING: A STUDY IN THE EVOLUTION
OF PROCESS CONTROL

Ramchandran Jaikumar

FROM ART TO SCIENCE IN MANUFACTURING: THE EVOLUTION OF TECHNOLOGICAL KNOWLEDGE

Roger E. Bohn

Foundations and Trends® in
Technology, Information and Operations Management
Vol 1, No 1 (2005) 1-120
© 2005 Mrinalini Jaikumar

the essence of knowledge

From Filing and Fitting to Flexible Manufacturing: A Study in the Evolution of Process Control

Ramchandran Jaikumar

(1944-1998)

Abstract

A well-known but previously unpublished monograph by Ramchandran Jaikumar (1945–1998) on the nature and history of manufacturing. The development of mass manufacturing ranks as one of the most important contributions to human living conditions ever – of the same magnitude as agriculture and modern medicine. This monograph examines the development of manufacturing over several centuries, through the eyes of a single industry and company. Throughout this period, the key has been precise control of manufacturing processes, rather than production speed *per se*. There have been six revolutionary transformations in manufacturing, each involving a shift in hard technology, the nature of human work, and the nature of process control. Each epochal change was a step in the evolution of manufacturing from art to science. Each required a different ethos of process control and forced a new system of manufacture. Machines, the nature of work, and the nature of the organization all had to change in concert.

Preface

This issue of *Foundations and Trends™ in Technology, Information and Operations Management* presents a classic but previously unpublished monograph by Ramchandran 'Jai' Jaikumar (1944–1998) on the history of manufacturing. The development of mass manufacturing ranks as one of the most important contributions to human welfare ever – of the same magnitude as agriculture and modern medicine. Many authors have addressed seminal changes in manufacturing history, such as the Industrial Revolutions, but this monograph takes a longer perspective. It follows the development of manufacturing from the Renaissance to 1985, and shows how manufacturing underwent multiple conceptual transformations, in which changes in technology led to shifts in the nature of work itself. These epochal transformations are emphasized by following the progress of a single industry – firearms – and single company – Beretta – over the entire period. Since the essence of the product changed little over the entire period studied – a chemical explosive pushes a projectile through a metal cylinder – firearms manufacture is an unusually clear opportunity to study changes in hard and soft manufacturing technologies. The most far-reaching changes were in process control, from the use of dimensional measurements around 1800 to the introduction of unmanned machining around 1980.

Each such shift required new ways of organizing work and a different ethos of management. Machinery, organization, scale, product line, and many other factors all had to change in concert to properly exploit the new concepts. And each new epoch represented an intellectual watershed in how people thought about manufacturing.

Prof. Jaikumar wrote the original monograph in the late 1980s while he was on the faculty of the Harvard Business School. [21] Although in that pre-Internet era it was available only as a hard-copy working paper, it became widely known and cited. Professor Jaikumar intended to publish it eventually, paired with a similar longitudinal examination of a continuous process industry. But other projects intervened, and it was never published. Professor Jaikumar died tragically in 1998, leaving behind a legacy of published and unpublished research. When Professor Uday Karmarkar of UCLA approached me for contributions to his new journal, I immediately suggested this piece.

I have made few changes to the main text – primarily clarifications. I have not attempted to incorporate research on manufacturing history done in the last 15 years, and the results are inevitably incomplete. I apologize for the errors and omissions. In partial recompense, I solicit comments and supplements to this monograph, and will undertake to add them to the Web version. I am especially interested in short essays that comment on the evolution of manufacturing in the last 20 years. For example, is the final epoch in the text, the Computer Integrated Manufacturing/FMS epoch, still the last word, or can we distinguish a new epoch, one based on computer networking? How should we think about process control extending across entire supply chains?

In conjunction with Prof. Jaikumar's original monograph I have written a new paper developing one of his themes in more detail: the transformation from art to science in manufacturing. [7] By taking advantage of concepts we developed jointly subsequent to his original monograph, I attempt a more precise and thorough treatment of this topic. Our hypothesis was that the shift from art towards science corresponds to changes in both knowledge about and process control of the physical technology. We developed a framework for describing technological knowledge that makes it possible to track changes in knowledge in great detail, identify gaps in knowledge, and describe trajectories of change. Firearms manufacture provides an excellent case study for testing these ideas using historical evidence. This paper will be published in a separate issue of *Foundations and Trends*; they will be merged in the book version. I have also included a short biography of Ramchandran Jaikumar, who had a unique range of interests and passions.

The passage of time and my own ignorance make it impossible to thank everyone who contributed to this research, but I know Jai would have singled out a few in particular. Beretta's management made this unique longitudinal research possible by providing assistance and access to the company archives. John Simon, who edited many of Jai's works, provided critical assistance with writing and research of both the original monograph and this version. The Harvard Business School and its Division of Research provided financial support. Baker Library and the Library of Congress provided access to rare illustrations from the 18th century. And Jai's wife Mrinalini and sons Nikhil and Arjun provided constant support. My own thanks to Uday Karmarkar and Zachary Rolnik for their support of this project, and to the Alfred P. Sloan Foundation for financial support.

Roger E. Bohn
San Diego, California
February 2005

1

Introduction[1]

Process control is the coordination of machines, human labor, and the organization of work to effect the manufacture of a product. It involves the specification and monitoring of machine setups and operating parameters, formulation of rules and procedures to govern operator–machine interactions, and decisions about the utilization of, and sequencing of, operations on a line. Although the details of process control can be quite different in different industries, a common theme that emerges from its study is the *evolution of manufacturing from an art to a science*. Inasmuch as the long-term viability and manufacturing competence of a firm is intrinsically tied to how one manages this evolution, it is important to understand the factors that drive it.

Manufacturing technology is, in essence, the technology of process control. Because one finds in the metalworking industry a great variety of processes being practiced at any time, and because the industry is large and has a long history, it is a useful base from which to study evolving patterns of process control in the mosaic of machines, labor, and the organization of work. Because aggregate data at the level of

---

[1] This monograph by Professor Ramchandran Jaikumar is being published posthumously. Details are provided in the Preface – Roger Bohn, editor.

the industry does not lend sufficient relief to the shifts in this picture, we take, as our unit of analysis, a single firm and category of products.

Within the firm we study the evolution of process control from the perspective of the work station – the locus at which technology and work come together and manufacturing takes place. Because we are interested in a particular aspect of technology and work, namely manufacturing's shift from art to science, we also examine the thinking behind the ideas that have shaped process control and the cognitive components of work.

We focus specifically on the segment of the metal fabricating industry engaged in the manufacture of firearms. A number of major manufacturing innovations have had their seeds in this industry: development of machine tools at the Woolwich Arsenal; interchangeability of parts at the Whitney and Colt factories; Taylorism at the Watertown Arsenal. Considerable scholarship has been devoted to the study of this industry, and we are also aided by the existence of a single firm, Beretta (*Fabbrica D'armi Pietro Beretta SpA*), whose history includes the assimilation of each of these manufacturing innovations.

Based in the city of Gardone in what is now northern Italy, and controlled by the same family for fourteen generations since 1492, Beretta has been engaged in the manufacture of firearms for five hundred years. Whereas functionally the product has remained much the same, and manufacturing is still based on fabricating precise metal parts, the detailed processes by which it is manufactured have changed considerably over time. Thus, the firm provides as ideal a natural experiment as one could have. Although it originated none of the major metal fabricating innovations, Beretta was quick to adopt every one of them.

To illustrate how the transformation in manufacturing technology has come about, we visit the arsenals in which the various innovations originated – the Woolwich Arsenal in England and the Colt factory and Watertown Arsenal in the United States – and review the works of the originators. What these individuals thought about and did is the story of the evolution of process control in the metalworking industry.

1.1. The Case for "Epochal" Change in Manufacturing

It will become apparent as the story unfolds that process control has evolved in a succession of epochs, each characterized by a fundamental shift, or "revolution," in manufacturing technology, the organization of work, and the nature of the firm. The story is related from the perspective of the individual at a machine, where process control is effected and the changes can be seen most vividly.

Six epochs of manufacturing process control can be delineated, preceded by a pre-manufacturing epoch in which products were made but not manufactured.

(1) The **Craft System** (circa 1500)

(2) The invention of machine tools and the **English System of Manufacture** (circa 1800)

(3) Special purpose machine tools and interchangeability of components in the **American System of Manufacture** (circa 1830)

(4) Scientific Management and the engineering of work in the **Taylor System** (circa 1900)

(5) **Statistical process control** (SPC) in an increasingly dynamic manufacturing environment (circa 1950)

(6) Information processing and the era of **Numerical Control** (NC, circa 1965)

(7) Flexible manufacturing and **Computer-Integrated Manufacturing** (CIM/FMS, circa 1985)

The first change in the technology of manufacturing firearms came some 300 years after Beretta started making guns. It was the English System of Manufacture, which was introduced at Beretta after the Napoleonic conquest of the Venetian Republic and the establishment of a state-run arms factory near Beretta's location. Much of our understanding of how the English System changed the nature of work comes from a visit to the shop of Henry Maudslay. Sufficient records of this founder of the machine-tool industry exist to form a picture of workshops of the late 18th and early 19th centuries.

The next era, the "American System," is illuminated by a visit to the Colt Armory. It brought to a high state of refinement a system of manufacture based on the notion of interchangeability of parts and the development and use of special purpose machinery. This system was showcased at the Crystal Palace Exhibition in 1851, and within 20 years had been adopted in whole or in part by most of the armories in Europe. Beretta adopted the entire system, contracting with the American firm Pratt and Whitney to build a complete factory at its headquarters in Gardone. The third epoch was the Taylor System, which perhaps even more than the first two revolutionized manufacturing far beyond the firearms industry. Taylorism was the basis of the vast expansion in firearms and other metalworking during World War II. Because company records at Beretta are incomplete for this period, we turn to Hugh Aitken's detailed explication of the introduction of the Taylor System at the Watertown (Massachusetts) Arsenal around 1900.

The first three epochs – those characterized by the English, American, and Taylor systems of manufacturing – related to the material world of mechanization. Each saw the manufacturing world as a place of increasing efficiency and control, substitution of capital for labor, and progress through economies of scale. These objectives were obtained through an engineering focus on machines and what could be done with them. The role of labor was increasingly seen as one of adapting to the machines and the contingencies of the environment – ultimately, of being yet another machine. Concurrently, the machines themselves became more elaborate, capable of ever greater precision and control. Underlying these developments was the principle of *increasing mechanical constraint.*

Abbot Usher, a historian of technology, observes that

> some of the impressive improvement of machines consists of refinement of design and execution. The parts of the machine are more and more elaborately connected so that the possibility of any but the desired motion is progressively eliminated. As the process of constraint becomes more complete, the machine becomes

more perfect mechanically ... The general line of advance takes the form of substitution of the more intense for the less intense forces, grading up through a long sequence that begins with types of human muscular activity ... There is a steady increase in potential (energy): we have to deal with a transition for machinery worked at a very low potential to machinery run at very high potential. The change in potential itself requires more and more careful constraint of motion because these highly intense concentrations of energy could not be applied to mechanisms until adequate control was possible. [34, p 116]

This world of mechanization reached its zenith in the 1950s. Already one could hear rumblings of a brave new world. In 1946 Brown and Leaver laid out, in a *Fortune* magazine article entitled "Machines Without Men," a blueprint for a new industrial order.[2] They had made the intellectual leap from mechanization to information processing. Norbert Weiner, in his prescient analysis of the power of information processing, gave credence to Brown and Leaver's world-view. Though it would be another forty years before we would see the first automated, workerless factories, the seeds for the emergence of a new paradigm were planted.

It is appropriate that James Bright completed his landmark study, *Automation and Management*, in 1958, for that year marks the end of the era of mechanization. Bright observed that

the average manufacturing system of 1956 ... can be regarded as no more than a crude assemblage of unintegrated bits of mechanism. These mechanisms themselves may reflect the utmost in the mechanical art of our times. Still, when collected under one roof and directed toward a particular production end, they are anything but a machine-like whole.

A hundred years from now the average factory of our day may be regarded as having been no different in philosophical concept from the factory of 1850 ... (Process) "design" has meant

---

[2] Cited in [29, p 68-70].

the collection of equipment for a production sequence – not the synthesis of a master machine. [8, p 16]

The glue that makes a collection of machines a manufacturing system is people processing information. The lack of integration Bright speaks of, and the intelligence needed to make machines function, were the focus of the three post-War epochs. The fourth epoch – the Statistical Process Control era – began in the 1930s in the electrical equipment industry, but in the 1950s Beretta was a leader in its implementation in arms manufacture. The fifth epoch grew out of numerical control, while the sixth and final epoch is the world of computer-integrated manufacturing and flexible manufacturing systems. Beretta was an enthusiastic adopter of all three and the discussion of these epochs therefore focuses on its experiences.

Collectively, these three epochs constitute a fundamental shift in the paradigm of production – from a world-view of managing *material transformation* to one of managing *intelligence*. This shift heralds a radical departure in the way we conceive of manufacturing. It is in promoting an understanding of the nature and impact of this transformation that this paper makes its principal contribution. In the dynamic world characterized by statistical process control, numerical control, and computer integrated manufacturing, we see a reversal of the trends of mechanization: increasing versatility and intelligence; substitution of intelligence for capital; and economies of scope rather than scale. Machines are increasingly seen as extensions of the mind meant to enhance the cognitive capabilities of the human being.

1.2. The Long View

An incontrovertible trend we see through the six epochs of process control is the evolution of manufacturing from an art to a science. As we shall see, each epoch represented an attempt to achieve a particular goal in the management of system variation, namely: accuracy, precision, reproducibility, stability, versatility, and adaptability. In the early epochs Beretta and its industry developed measures of the product, then gained control of the process. Next it mastered variability, first in the machine, then in the human. Finally, it studied, and then controlled, contingencies in the process until it was able to extract general principles and technologies that apply in a variety of domains. In short, it achieved *versatility*. It will become apparent that the ethos of process control required to manage each of these is quite different. It is extremely difficult for a firm to manage the conflicting demands of two successive process control paradigms. Therefore, the management of technology required a quick transition from one to the other.

There is a consistency in these six epoch shifts as they were experienced by Beretta.

- Each epochal change represented an intellectual watershed as to how people thought about manufacturing and its key activities.

- Each epoch entailed the introduction of a new system of manufacture; machines, the nature of work, and the organization all had to change in concert to meet a new technological challenge.

- The technological change of each epoch focused on the solution of a new process control problem, but in all six cases this problem revolved around controlling variation.

- Most of the gains in productivity, quality, and process control achieved by Beretta over its 500-year history were realized during the assimilation of the six epochal changes and very little in between.

- It took about ten years to assimilate the change incurred by each epoch.

- All of the changes were triggered by technology developed outside the firm.

Clearly, each of these epochal changes could affect all metal fabricating industries, and they did. But by examining these changes at the level of the work station in a single firm concerned with the manufacture of a single type of product, the firearm, we can see their impact in sharpest relief and observe a consistency that suggests powerful lessons for the management of technology. Our objective in scrutinizing a variety of historical records is not to trace the origin of ideas in process control, or even the full impact of those ideas on manufacturing, but rather to analyze how they have changed the *nature of manufacturing*, effectively moving manufacturing from an art towards a science.

Table 1.1 summarizes some of our findings about the six epochs along dimensions that provide insight into the nature of these epochal shifts.

1.3. Plan of the Monograph

The balance of this introduction discusses the fundamental technical problem of manufacturing, namely controlling the variation inherent in any physical process. Section 2 describes the way firearms were made before the development of manufacturing, by individual master craftsmen. Sections 3 through 8 describe the six manufacturing epochs, in chronological sequence. Beretta's own experiences are discussed, as well as the historical origins of each epoch. In each case, the radical nature of the transformation from the previous epoch is emphasized.

Section 9 concludes the monograph by pointing out how in some ways the nature of manufacturing today resembles that of 200 years ago. A few dozen expert workers with high discretion produce a wide variety of products. Yet other aspects have changed beyond recognition. Human muscle power is irrelevant, output per worker is up 500-fold, and rework is virtually zero.

This monograph is intended to be read in conjunction with additional material. The Preface introduces the monograph and explains its origins. A biography of the author appears at the end. A companion article extends the theme of "manufacturing moving from art to science." It provides a precise model of what this means. The level of detail of technological knowledge increases over time, approaching but never reaching comprehensive scientific "first principles" models of all key phenomena. This permits more decisions to be made according to programmed procedures, without human discretion, as described in the current monograph. These two dimensions, of knowledge and control, tend to grow in concert. When technologically disruptive innovations arrive, however, they step backwards because the detailed knowledge

	English System	American System	Taylor System	Statistical Process Control	Numerical Control (NC)	Computer Integrated Manufacturing
Size Trends — Introduced at Beretta (world)	1810 (1800)	1860 (1830)	1928 (1900)	1950 (1930)	1976 (1960)	1987
# of People (Min. Scale)	40	150	300	300	100	30
Number of Machines	3	50	150	150	50	30
Productivity Increase *	4:1	3:1	3:1	3:2	3:1	3:1
Number of Products	Infinite	3	10	15	100	Infinite
Nature of work — Standards for Work	Absolute product	Relative product	Work standards	Process standards	Functional standards	Technology standards
Work Ethos	"Perfection"	"Satisfice"	"Reproduce"	"Monitor"	"Control"	"Develop"
Worker Skills Required	Mechanical craft	Repetitive	Repetitive	Diagnostic	Experimental	Learn/ generalize/ abstract
Control of Work	Inspection of work	Tight supervision of work	Loose of work/ tight of contingencies	Loose supervision of contingencies	No supervision of work	No supervision of work
Organizational Change	Break-up of guilds	Staff-line separation	Functional specialization	Problem-solving teams	Cellular control	Product/Process/ Program
Staff/Line Ratio	0:40	20:130	60:240	100:200	50:50	20:10
Line Workers per Machine	15	3	1.6	1.3	1	0.3
Technology Keys — Process Focus	Accuracy	Precision: Repeatability (of machines)	Precision: Reproducibility (of processes)	Precision: Stability (over time)	Adaptability	Versatility
Focus of Control	Product functionality	Product conformance	Process conformance	Process capability	Product/ process integration	Process intelligence
Instrument of Control	Micrometer	Go/No-Go gauges	Stop watch	Control chart	Electronic gauges	Professional workstations
Rework**	.8	.5	.25	.08	.02	.005

* Over previous epoch **As fraction of total work

Table 1.1 Summary of manufacturing epochs

underlying the innovation must be developed. The current state of knowledge and its evolution can be captured by directed graphs showing current knowledge about the technology.

1.4. Control of Variation

Historical studies of manufacturing evolution typically emphasize issues related to scale and energy, such as the evolution of power sources from human/animal power, to water power, to steam, to electricity. Superficially, the increasing intensity of energy use enabled the long-term manufacturing trends of increasing speed and scale – machines that go faster and make more at a time. But in technological terms power is secondary. Crudely applying more force has been feasible, at least since the invention of steam engines, just by building bigger mechanisms. But with no corresponding progress in control, a bigger process will only make junk more rapidly. Therefore the key to progress has been gaining better *control* over manufacturing processes, which permits simultaneous increases in both precision and force.

Consider a product that comprises two or more metal components that must be joined together. The manufacture of such a product entails two types of processes: *fabrication* processes by which individual components are formed, and *assembly* processes that marry the discrete components into subsystems and a final system. For our purposes, it is sufficient to say of the latter that it comprises a sequence of operations whereby the constituent pieces of a product are selected, located, fitted, and bonded. It is with the former set of processes – those that govern metal fabrication – that we are primarily concerned in tracing the evolution of process control.

The purpose of a metal-fabricating process is to create, according to precisely prescribed specifications, the form, physical characteristics, and finish of a metal component (part). The process is executed by a set of people, machines, and procedures, and a measure of their effectiveness is the ability to produce correct and specific parts. Inasmuch as a process never performs identically each time, some variation in the parts produced is inevitable. Sources of variation lie in people, machines, and procedures, as well as in the object being fabricated. A measure of the effectiveness of process control is the degree to which variation is minimized. The goal of process control is to limit such variation, and *the study of process control is the study of the kinds of variation that can occur, their sources, and the means by which they can be managed.*

Proper functioning of the finished product depends on multiple characteristics of each component, such as physical dimensions, strength, and surface finish. The desired level of each characteristic is its *target specification.* For example, consider two metal parts which are intended to fit together by having a cylindrical peg on one part that fits into a round hole on the other, such that they can rotate relative to each other. Obviously, the diameter of the peg must be no greater than that of the hole, else they won't mate. But if the peg is too much smaller than the hole the parts will rattle against each other, and the mechanism will work poorly or not at all. The product designer deals with this by specifying a target diameter for the peg and the hole such that they will fit properly, and a range of allowable variation around the targets. Such targets are called *specifications*, and the ranges are called *tolerances.*

But the realized characteristics of components produced by a particular process are not identical to the target specification or to each other, so their behavior must be described by frequency distributions. The difference between the achieved mean dimension and the target specification is the *accuracy* of the process. The variation of the distribution around its mean tells us to what degree the process is capable of achieving the desired performance; the smaller the dispersion around the process mean the more capable the process. The reciprocal of the variance is the process *precision*, which measures the ability of a machine to execute identical performances and the ability of people and procedures to direct the machine.

Variation arises from a multitude of sources. To overcome variance attributable to machines we strive for *repeatability*; to overcome variance attributable to people and procedures we strive for *reproducibility.* If we measure, for a single component and dimension, the means for sequential lots we would find that over time the mean of the process changes. The standard deviation over time of the process mean, defined as the *stability* of the process, is a measure of how well it performs over time. *System variance* is the net variance due to accuracy, repeatability, reproducibility, and stability.

The above measures of variance assume that we have not made any adjustments to the process. In practice, we always make adjustments to a process when something goes wrong and a process that accommodates such adjustments is obviously desirable. Accuracy, as noted earlier, is the systematic bias in a process, stability the manner in which that bias shifts over time. To the extent that we can adjust the process we can correct the bias and bring it closer to the desired standard. The capability of a process to make dynamic adjustments and correct for bias is termed *adaptability*.

The requirement for adaptability is quite different depending on whether we want to make one component or a large number of identical components. To be adaptable with a sample of a single component a process must have a high degree of accuracy. More important in a process for producing large quantities are precision and stability, as we can almost always compensate for inaccuracy by making adjustments. The greater the stability of a process, the less frequently it will have to be adjusted.

Before proceeding with our discussion of the evolution of process control we need to define a further notion, that of *versatility*. Versatility is the ability of a process to accommodate variety in product specifications. It is quite different from the notions discussed above, yet it has important implications for process control. As greater versatility usually reflects greater complexity in a task, the sources of variation can be expected to increase when versatility increases, if no other changes are made.

Process control is central to manufacturing because better control reduces variation, which enables a number of benefits: higher production rates, lower rework, tighter tolerances, and less raw material. In turn these improve characteristics that end-users care about: cost, product variety, and product quality. For firearms, product quality measures enhanced by reduced variation include weight, power, durability, and shooting accuracy. The benefits are taken as some combination of these attributes depending on market preferences, such as the orders of magnitude improvements in both firearms manufacturing productivity and product performance over two centuries. In the end a seemingly manufacturing-specific issue, process variability and its control, is at the center of the technological and economic revolutions of the last centuries.

2

Gun-making in Gardone – The Craft System

For hundreds of years after its inception, gun-making in Gardone, Italy, changed little. By contrasting the practices related below to those described subsequently in connection with the English and American systems, we can begin to understand the scope of the changes with which Gardone's gunsmiths had to cope.

The locking mechanisms forged in the shops of Gardone gunmakers in the 1780s were little changed from those of 300 years earlier. Figure 2.1 and Figure 2.2, taken from *Diderot's Encyclopedia*, illustrate the nature of the shops and kinds of tools and measuring instruments then in use. Although the shop depicted in the plates did assembly, shops that fabricated components would not have looked much different. There would be a forge to make small components and a crude drilling machine, but there would be no planer machines to do metal cutting. Hammers, chisels, and files were the principal tools, calipers and wooden rules the only measuring devices. Human muscle supplied the mechanical power.

Shops kept models of locking mechanisms from which the craftsmen worked, constantly comparing the component being manufactured with the model. Components were hand-forged, filed to shape, fitted together, and then hardened. The bulk of the work in these shops consisted in

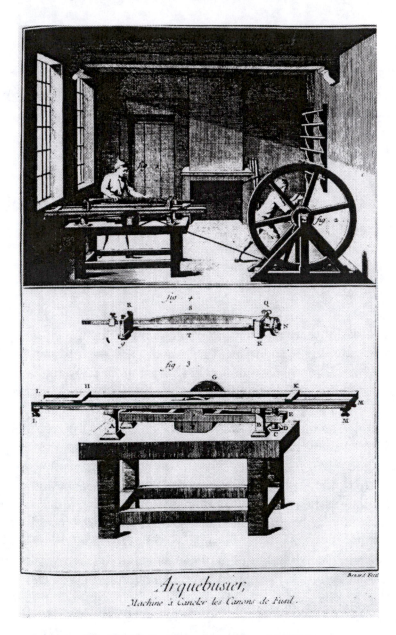

Fig. 2.1 Craft epoch gun-making shop [10]

filing and fitting pieces. The assembly process was imprecise, a matter of repeated trial and error adjustment to get pieces to fit – essentially

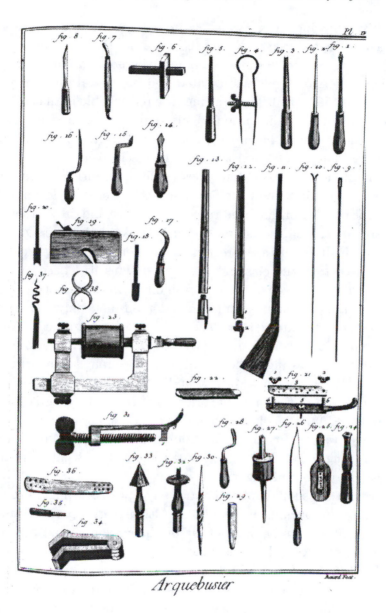

Fig. 2.2 Craft epoch tools [10]

100% rework.

Although models were far and away the primary means by which artisans communicated design intent, some designs were replicated in primitive drawings that were circulated among the masters. The engravings reproduced in Figure 2.3 are from the introductory plates of *Vershiede Stucke Fur Buchsenmacher* by Johann Christoff Weigel, probably the most widely circulated and influential gun design book of the span 1650–1750. The drawings are remarkable in that they carry no specifications or dimensions. Only design intent and functionality are communicated, the interpretation of the design by the master serving as the basis for constructing the mechanism.

International distribution of designs for gunsmithing dates to 1635, the year of publication of the first book of patterns by Phillipe Daubigny. The custom proliferated rapidly in France and, after about 1700, in Germany as well. It was the German books that exercised a strong influence in Italy's Brescia region. By the end of the first quarter of the 18th century the classical Brescian designs had been abandoned by gunmakers in favor of the new fashions then dominant in Germany and Austria. Brescian gunmakers adopted not only German gun architecture and external structure, but also German mechanisms.

Production involved the master, the model, and a set of calipers. If there were drawings, they indicated only rough proportions and functions of components. Masters and millwrights, being keenly aware of the function of the product, oriented their work towards proper fit and intended functionality. Fit among components was important and the master was the arbiter of fit. Apprentices learned from masters the craft of using tools. Control was a developed skill situated in the eyes and hands of the millwright.

A master's shop employed about eight people. Annual production was about 260 locking mechanisms. Although the pace of work was usually quite leisurely, the output of these shops could as much as quadruple during peaks of demand.

In contrast to gun barrel making shops, which were functionally focused and organized around five classes of workmen – forgers, borers, smoothers, filers, and finishers – shops engaged in the construction of locking mechanisms were product-focused. The work in the latter shops consisted in bringing the components together and obtaining the right fit. Everyone in the shop was involved in all five stages of the production process, which consisted of forging, filing, fitting, and polishing.

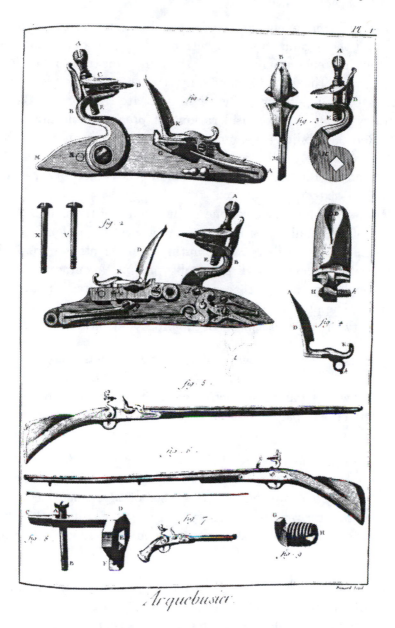

Fig. 2.3 Early gun design drawings [35]

As the principal activity, fitting, involved filing and fitting two or more components and polishing the composite workpiece, we see that the fabrication of components and their assembly were closely intertwined.

Given the organization and activity of these shops, what can we say about system variance? Note that the construction of locking mechanisms at this time involved only the use of hand tools and vises. There were no jigs to properly align or locate components. With no machinery to speak of, considerations of precision and stability are moot. Reproducibility accounted for all system variance, which was very high. With only calipers and wooden scales, and control completely in the hands of the craftsman, the standard deviation of error was as large as one-sixteenth of an inch.

With such high variance, one cannot think of the manufacture of a batch of nominally identical items together, only of making each individual item. And fit between mating components is impossible to achieve without having both physically present. Accuracy is achieved here through adaptability, that is, the ability of the craftsman to adjust the contours appropriately.

Note two important aspects of the process we have been examining.

- First, an assemblage of diverse components was required to fabricate and assemble a single product. The craftsman had to view the parts for each firearm independently of the same functional part for the next firearm. The concept of "identical parts" did not exist.

- Second, the measure of skill lay in degree of adaptability, that is, the ability of the craftsman, or operator, to adjust to a wide variety of conditions and the speed of adjustment necessary to obtain the required accuracy. The speed of adjustment between high-skill and low-skill workers could be as great as four to one.

The adaptability of the operator being so important, it was only natural that managerial response was directed towards improving skills and maintaining a skilled work force. Systems that developed adaptive skills flourished and the master–journeyman model survived for many centuries.

Inasmuch as adaptive skills are really contingent responses to a wide variety of work conditions, procedures cannot readily be transferred. Critical knowledge was mainly tacit, and a journeyman had to learn by observing the master's idiosyncratic behaviors. The master, who could solve the most difficult of problems, fashioned each product such that quality was inherent in its fit, finish, and functionality.

It should be noted that adaptability by craftsmen is needed because of the inability of a process to obtain adequate accuracy, precision, reproducibility, and stability. Thus, it is a response to a deeper problem. Fundamental process improvement that reduces system variance would reduce the need for adaptability, and thus the very skills of the master. But to reduce system variance below the craft system, it would be necessary to:

- devise tools that would lend greater control and, thus, precision to the metal cutting process;

- introduce more accurate measuring instruments so that one could obtain constant feedback on the state of the product and thereby strengthen adaptive response;

- simplify product designs to reduce variance associated with reproducibility, i.e., to allow different people to make a part in the same way.

A fundamental shift in the focus of technological attention is inherent in all these requirements.

3

The English System of Manufacture

The first change in the technology of manufacturing firearms occurred some 300 years after Beretta began making guns. It came in the form of the English System of Manufacture, which was introduced at Beretta as a result of the Napoleonic conquest of the Venetian Republic and the establishment of a state-run arms factory at Brescia, the capital of the province around Gardone (c. 1800).

The machine tool industry was born in England in the late 18th and early 19th centuries through the agency of English mechanics who devised tools that added greater precision to the process of metal cutting and introduced accurate measuring instruments that helped them achieve a high class of workmanship. The building and use of tools was the focus of their attention. The tools themselves, being general purpose, could be used to fabricate a variety of workpieces. The apprentices who trained in the shops of the great English mechanics were much sought after, having become skilled in the use of instruments and machines. Their skills being applicable to the building of many different workpieces, apprentices focused on the *tools* they used rather than on the *products* they fabricated.

With the development of machine tools, the functionality of a product need no longer be viewed together with the process used to make it. The process took on a life of its own, enabling process improvements to be made independently of product constraints. This was the intellectual leap that freed the development of technology from the constraints of the product. Once it occurred, the flowering of technology was rapid. Within 50 years the technological landscape was revolutionized.

The seeds of the new system of manufacture that would utilize the new technology were sown by a young mechanic, Henry Maudslay (1771–1831), who worked at the Woolwich Arsenal. Much of our understanding of how the nature of work changed as a consequence of the introduction of the English System derives from a description of Maudslay's shop; sufficient records of the work of this founder of the machine tool industry survive to enable us to paint a picture of what the workshop of the late 18th and early 19th centuries looked like.

The effect of the English system on Beretta, when it was implemented around 1810, are summarized in Table 3.1. Products were still infinitely varied as before, and all employees worked with their hands on the actual objects being made (no separate staff activities), but in other respects it marked the birth of what we now know as process control.

3.1. Tools for the Woolwich Arsenal

The tools being built by Maudslay in the 1790s were a source of great wonder to his fellow workers. A born craftsman whose skill was the pride of the entire shop, Maudslay supplemented dexterity with an intuitive power of mechanical analysis and a sense of proportion possessed by few men. He exhibited a genius for accomplishing his ends by the simplest and most direct means.

Of all his phenomenal inventions, Maudslay is best known for the development of the slide rest and its combination with a lead screw operated by change gears (Figure 3.1). One of the great inventions of history, it is still used in almost every machine tool.

Like most great inventions, the slide rest was a product of many minds. Leonardo da Vinci had made crude drawings of it. Besson's screw cutting lathe, built in 1569, shows a lead screw. *Diderot's*

Summary of Epoch		English System
Size Trends	Introduced at Beretta (world)	1810 (1800)
	# of people (Min. Scale)	40
	Number of Machines	3
	Productivity Increase (over previous epoch)	4:1
	Number of Products	Infinite
Nature of work	Standards for Work	Absolute product
	Work Ethos	"Perfection"
	Worker Skills Required	Mechanical craft
	Control of Work	Inspection of work
	Organizational Change	Break-up of guilds
	Staff/Line Ratio	0:40
	Line workers per machine	15
Techno-logy Keys	Process Focus	Accuracy
	Focus of Control	Product functionality
	Instrument of Control	Micrometer
	Rework (as fraction of total work)	.8

Table 3.1 Effects of English System at Beretta

Encyclopedia shows an early slide rest. Samuel Bentham anticipated the combination of slide rest and lead screw operated by change gears. [30, p 28] "When the motion is of a rotative kind," Bentham wrote in his 1793 patent, "advancement [of the tool] may be provided by hand, yet regularity may be more effectually insured by the aid of mechanism. For this purpose one expedient is the connecting, for instance, by cogged wheels, of the advancing motion of the piece with the rotative motion of the tool." (British Patent 1951, April 23, 1793) But it is to Maudslay that the distinction of actually designing and developing the first power-driven and controlled lathe belongs.

To take the place that it did in industry, the lathe had to possess a number of features, enumerated by Robert Woodbury below, which Maudslay was able to synthesize.

An industrial lathe must have: first, the ability to machine an iron or steel workpiece of a substantial industrial size. In order to meet this requirement, the lathe must itself normally be made of iron

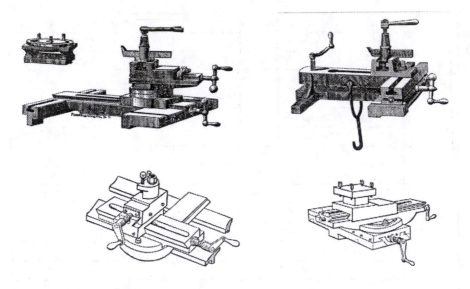

Fig. 3.1 Slide rest, circa 1885 [17]

or steel and have its various parts of dimensions such that it can *withstand the stresses* set up in it by cutting the ferrous metals.

Second, the industrial lathe must also be supplied with a *source of power* and means of its transmission to the workpiece and to the cutting tool adequate for cutting iron and steel at rates which are economical. This requires a suitable headstock spindle with means for its drive, and a tool carriage with its feed.

Third, the industrial lathe must itself be constructed with *adequate rigidity* and *precision* so that it is capable of producing a precision nearly equal to its own in the workpieces turned upon it ... Rigidity in a lathe is provided partly by the material of which it is made and partly by the design of its parts, but precision depends also upon the accurate construction of certain of its features, especially the spindle bearings, the guideways, and the *lead screw*. The precision actually needed in the industrial lathe at any given period is somewhat greater than that required for the work to be done on it.

Fourth, the industrial lathe must have *flexibility*. Only a few machine shops in the mid-19th century could afford to have specialized machine tools, such as a boring engine, a screw-cutting machine, or a gear-cutting machine. Most shops had to depend upon a lathe, a planer or shaper, and a drilling machine ... To achieve flexibility the lathe needs at least *change gears* for both screw cutting and longitudinal feed of the tool, cone pulleys or some other means of varying the speed of the workpiece and the cutting rate, a sliding tailstock to take work of different lengths, and a chuck or a face plate for boring or for other turning not possible with the workpiece mounted between centers. [37, pp 96-97. Italics added]

The machine that Maudslay built in 1800 (Figure 3.2) was, according to Roe, "distinctly modern in appearance. It has a substantial, well-designed, cast-iron bed, a lead screw with 30 threads to the inch, a back rest for steadying the work, and was fitted with 28 change wheels with teeth varying in number from 15 to 50." [30, p 104]

The lathe of 1800, however, was the beginning rather than the end of Maudsley's work on the screw. In the course of the next 10 years he made exhaustive studies of the problem of screw cutting and succeeded in placing this fundamental aspect of metalworking upon a solid foundation ... Every resource was exhausted in the development of accurate original screws. Beginning with the best of the hand methods, numbers of screws were prepared and the best of them selected for further work in specially constructed lathes. "A very excellent brass screw about 7 feet long" was finally constructed, "which was less than one-sixteenth of an inch false in its nominal length." A device was then constructed to remedy this error and the new screw produced was examined with micrometric apparatus ... [It and another screw] were then subjected to further corrections until they became accurate within any margins of error then significant for mechanical or even scientific purposes. [34, p 369]

Upon such precision lathes Maudslay cut some of the best lead screws to that time. One of these "was principally used for dividing scales for astronomical and other metrical purposes of the highest class.

Fig. 3.2 Maudsley's lathe [37]

By its means divisions were produced with such minuteness that they could only be made visual by a microscope." [30, p 41] "I believe it may be fairly advanced," wrote Holtzapffel, "that during the period from 1800 to 1810, Mr. Maudslay effected nearly the entire change from the old, imperfect, accidental practice of screw making to the modern, exact, systematic mode now generally followed by engineers." [16 p 647]

His many inventions notwithstanding, Maudslay's importance lay less in the development of machines than in the founding of the machine tool industry and the radical transformation of shop floor practice. "Maudslay's standard of accuracy," Roe observes, "carried him beyond the use of calipers." In his workshop, Maudslay kept a highly accurate bench micrometer, which he referred to as "The Lord Chancellor." About sixteen inches long, the micrometer had two plane jaws and a horizontal screw, a scale graduated in inches and tenths of an inch, and an index disk on the screw graduated to one hundred equal parts.

"Not only absolute measure could be obtained by this means," remarked James Nasmyth, "but also the amount of minute differences could be ascertained with a degree of exactness that went quite beyond all the requirements of engineering mechanism; such, for instance, as the thousandth part of an inch." [28, p 150]

Nasmyth further observed that "the importance of having Standard Planes caused him [Maudslay] to have many of them placed on the benches beside his workmen, by means of which they might at once conveniently test their work ... This art of producing absolutely plane surfaces is, I believe, a very old mechanical 'dodge.' But, as employed by Maudslay's men, it greatly contributed to the improvement of the work turned out. It was used ... wherever absolute true plane surfaces were essential to the attainment of the best results, not only in the machinery turned out, but in educating the tastes of his men towards first-class workmanship."

Whitworth could later assert "the vast importance of attending to the great elements in constructive mechanics – namely, a true plane and the power of measurement. The latter cannot be attained without the former, which is, therefore of primary importance ... All excellence in workmanship depends upon it." [36, p 125]

This striving for accuracy and workmanship was another of Maudslay's lasting legacies. Through his workshop, which employed several hundred men at one time, passed nearly the entire coterie of great machine tool builders. Clement, Roberts, Whitworth, Nasmyth, Seaward, Muir, and Lewis showed throughout their lives and in a marked way Maudslay's influence upon them. The methods and standards of Maudslay and Field, spread by the former's workmen into the various shops of England, made world leaders of English tool builders.

Under the leadership of the "Maudslay men" all of the great metal-working machine tools achieved a form that remained essentially unchanged for nearly a century. England enjoyed unquestioned leadership in the machine tool industry, supplying nearly all of the machine tools used in France and Germany, whose own machine tool development lagged a generation or two behind.

The influence of the English mechanics was not limited to the building of machinery. Their mode of apprenticeship had produced a cadre of individuals who conceived a new system of manufacture, the foundation of which was mechanical engineering and the roots of which lay in principles of measurement and accuracy and the ability to meet tolerances. Theirs was not a skill based on knowledge of the functions required to manufacture a specific product, but rather knowledge of tools and of scientific principles of measurement skills. These skills, and the general-purpose machine tools being produced at the time, could be applied to a variety of products. The accuracy, precision, and productivity of general-purpose machine tools was greater than that of hand tools and other product-specific tools then in use in the various industries. This system of manufacture quickly came to predominate in machine shops in England and soon spread throughout Europe.

3.2. The Engineering Drawing

The engineering drawing, as a medium of communication in engineering work, did not exist prior to 1800. Engineering work was defined by a physical model of a product that was to be reproduced. In the manufacture of a musket barrel, for example, a worker would ensure that the dimensions of the barrel on which he was working corresponded to those of a model barrel by using calipers to transfer measurements from one to the other. Because each worker needed his own model barrel to work from, the greater the number of workers a shop had, the greater the number of model barrels it had to supply. As it was impossible, given the standards of accuracy of the time, to make all model barrels identical, the manufactured barrels were all different.

La Geometrie descriptive, written by Gaspard Monge in 1798, was the first formal treatise on modern engineering drawings. In it, Monge (1746–1818) developed the theory of projecting views of an object onto three mutually perpendicular coordinate planes (such as are formed by the front, side, and top of a cube) and then revolving the horizontal (or top) and profile (or side) planes onto the same plane as the vertical (front) plane. The fundamental theory of all orthographic (mutually perpendicular) projection is derived from Monge's descriptive geometry.

Monge showed how drawings need to be dimensioned. Dimensions, the size specifications added to the shape description as provided by the orthographic drawings, consisted of the numerical values of the measurements directed to the proper location on the part and the relevant surfaces or locations on the object. For drawings to replace models as a medium of communication, one needed accurate measuring instruments. The English system of manufacture, with its basis in measure, created a variety of such instruments.

Together, mechanical drawings and the English system altered the organization of work. With an objective standard of performance (a mechanical drawing) that was the same for every worker models were no longer needed. Work could be compared to the desired standard using an objective measure of performance, the micrometer (Maudslay's "Lord Chancellor"). The master was no longer needed for guidance or approval; the worker could obtain the former from a drawing and verify his work with the appropriate measuring instrument.

With clear specifications of what is required and an objective standard with which to compare performance, we would expect system variance due to inaccuracy to be markedly reduced. With workmanship a prized objective and no longer product-specific (a workman trained to turn out a metal shaft for a horse-drawn carriage could now turn out a rifle barrel as well), we would expect the guild system to collapse. With workers no longer ingratiated to a master and free to leave as soon as they had developed the necessary skills, we would expect to see a market develop for skilled labor. We will see all of these things happen at Beretta. What the merchants of Brescia and Venice and the Doges managing the Venetian Arsenal could not do, the mechanical drawing and micrometer achieved in a span of fifteen years.

3.3. Gardone Shops for Barrel-Making

Gardone became part of the Napoleonic French Republic on 21 September 1792. As the revolution had deprived the French aristocracy of political and economic power, so the Napoleonic era had a profound effect on Gardone gunsmithing.

The cooperative artisan organization of Gardone, considered "antidemocratic" by the Francophile municipal authorities, had been completely dismantled. Anyone could now engage in the various professions, or "arts," of barrel making. The barrel-master under the guild structure was, under the English system, replaced by the machine operator. Under this system, each person was responsible for all aspects of making a component.

The English system of manufacture introduced to Beretta during the French occupation transformed the arms factories in nearby Brescia. The French had found that with drawings and general-purpose machines they could have one large factory instead of many small shops. The new, state-owned arms factory established in Brescia to maintain the supply of arms to the French army was furnished with the latest machinery, imported from France and operated under the factory system of production. For the first time, the masters had to contend with a radically different technology and organization of work. Marco Cominassi writes of the factory:

> The work force of the Imperial-Royal Factory in Gardone consists of 180 skilled workers, not counting apprentices or the women who work there; together they could produce two thousand barrels a month. A huge building, property of the public treasury, serves as the residence of the supervisors and agents, and of the captain when he comes here from Brescia; here, too, the iron that is advanced to the workers is given out, and where they bring the finished product. The barrels are proved in the presence of the captain.
>
> The work is divided up among five classes of craftsmen, called the forgers, the borers, the smoothers, the filers and the finishers. Each of these groups elects a leader who retains office for three years, and lives in Gardone without ceasing to participate in the work; these leaders, under the presidency of the captain, form the administrative council of the factory.
>
> The forgers receive their iron in flat rectangles purified under the drophammer; they wrap it around the mandril by force of fire and hammer so that the two long edges are fused together and form a barrel.

The barrel having been roughed out and certified as perfect by a supervisor, it passes on to the borers, who use water-driven machinery [see Figure 3.3] to clean out the scaly bore with rough circular files, first narrow, then of ever greater diameter, until the required size is reached. But since the borers cannot make the bore perfectly cylindrical with their instruments, the barrel passes next to the smoothers, who subject it to subtle and careful labours. The external finish is then entrusted to the filers who, with a diligence that is partly a specialty of this factory, reduce the barrel to final shape by bringing it into contact with a large water-powered sanding disc. Next comes the polishing phase, done by the finishers with special files, and finally the fine-polishing is done with various abrasives by the women. The breechers, those who fit the sights and the proofers constitute an appendix to the five classes of above-named craftsmen.[1]

The effect of the machinery and organizational structure of the state-owned factory at Brescia was like a shock to Gardone. The productivity of the factory and the quality of the muskets it produced for the military far exceeded that of the shops of the Gardonese artisans. Artisans at the state-owned factory were, with but three years of training, turning out a product far superior to that being turned out by masters who had devoted a lifetime to their art. Only the mercantile contracts of Beretta kept the masters in business at all. Their markets in the Levant were being lost to French, English, and Belgian competitors. If Beretta and the other shop owners were to remain viable, they would have to modify their activities. It would not be easy to assimilate the changes that had occurred in the industry, but the Gardonese had little choice but to try.

Pietro Antonio Beretta (1791–1853) attracted a number of newly trained artisans from the state factory. He purchased three new machine tools and expanded the scope of activities within his shop, which grew from eight to forty people and enjoyed a four-fold increase in productivity. The greater size and increased productivity of the rejuvenated shop

---

[1] From "Notes on the Arms Industry in Gardone in the Trompia Valley," *Giornale* (Imperial-Royal Lombard Institute for Science, Letters and Art). In [18, pp 199–202]

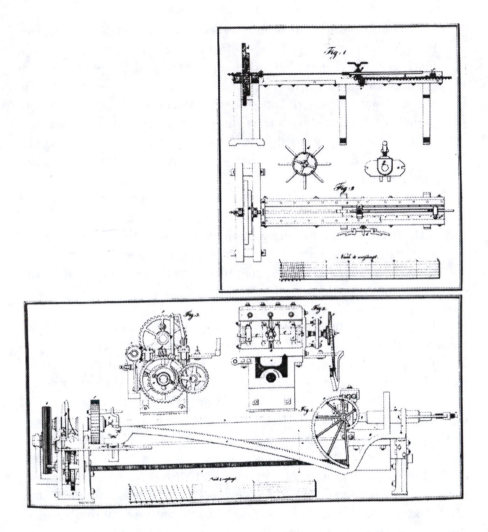

Fig. 3.3 Boring machine (top) and barrel lathe (bottom), circa 1829 [18, p 203]

represented an order of magnitude increase in productive capacity. In 1815, Beretta traveled widely, establishing contacts with importers, wholesalers, and retailers. The network he established was woven tighter by his son, Giuseppe, who achieved the unification, in a single factory, of the complete manufacture of a firearm.

The importance of the attention of the master to product quality gave way to the availability of operators trained under the new system.

It was the availability of the skills of the latter that became the new constraint on growth. This constraint would not be long-lived, though. It was soon to be relaxed by the introduction of the American System of Manufacture, which used special purpose machinery and tooling that required less skill.

A visit to the Colt Armory illuminates this next half-century of progress. The Colt factory brought to a high state of refinement a system of manufacture based on the notion of interchangeability of parts. After the famous Crystal Palace Exhibition in 1851, this "American System" was, over the course of the next two decades, adopted in whole or in part by most of the armories in Europe. Beretta adopted the entire system, contracting with the American firm Pratt and Whitney to build a complete factory in Gardone.

4

The American System of Manufacture

While the English were evolving a system of manufacture around the ethos of *accuracy*, a new system, based on *precision* and interchangeability of parts was being developed in the United States. The difference occurred because in the English System mechanics and engineers made parts to fit (i.e., to mate with one another) as closely as possible, while interchangeability, by contrast, relies on the existence of clearance between parts. In the English System, the better the fit, the better the workmanship, with "perfection" being the objective. As "fit" was achieved by concentrating on the relationship *between* components, one *made* parts for each subassembly one at a time.[1] The parts being assembled were then filed by hand until the mated surfaces fit tightly. The result is that each part and each subassembly are unique.

The greater the clearance between mating surfaces, the more likely parts will be interchangeable. Thus, the objective of interchangeable manufacture was to move from perfection of fit towards the greatest possible clearance, as long as the clearance was not too large to lose the functionality of the product. In doing so, the intellectual problem changed from generating perfection of fit by custom filing and fitting

---

[1] Paul Uselding discusses the differences between the two systems of manufacture and their relation to precision and accuracy in [33].

to managing clearances between components in large batches. These concerns are at opposite poles.

Clearances allowed for variance, and management of these variances were the hallmark of the American System of Manufacture. Interchangeable manufacture allowed for the separation not only of fabrication and assembly, but also of the different operations in fabrication from one another. Managing variances entailed prescribing limits and then achieving the precision imposed by these limits by developing (1) machinery that was constrained in its operation, and (2) a system of inspection based on gauges that would ensure that fabricated parts were, indeed, interchangeable.

The simultaneous introduction of special purpose machines and systems of gauging and inspection had the effect of reorienting the thinking of engineers away from making individual components towards the development of systems for manufacturing large lots of components. Charles Babbage, in his celebrated work, *On the Economy of Machinery and Manufacture*, was the first to distinguish the English and American systems on the basis of *making* versus *manufacturing*. [6] Engineering problems were radically different between the two systems. The essential feature of precision manufacture was exact duplication utilizing matched or common fixtures, tools, and size gauges. Workpieces were produced to fit these fixtures, tools, and gauges, rather than to exact size relative to a universal standard of measurement. Thus, the *accuracy* of parts, according to the English System's concept of deviation from engineering drawings, would generally be worse, yet because every part in the lot was *consistent*, they could be interchanged.

Although the first complete manufacturing system based on interchangeable parts, a system for making pulley blocks, was built by Brunel, Bentham, and Maudslay at Portsmouth in 1795, their achievement did not alter the intellectual ethos of technological achievement in England. Development of the system was left to the Americans. Our concern with interchangeability in America is not with its origins, which are the subject of some debate, but rather with its effects on the nature of work.

The effects of the American System at Beretta, which was introduced in 1860, are summarized in Table 4.1. Output per worker increased by a factor of 3 while number of workers quadrupled and the number of machines grew from 3 to 50. Products became highly standardized, with only three different products made in the factory.

4.1. The Whitney Factory

Eli Whitney, in carrying out a 1798 contract from the United States government for the manufacture of firearms, employed mainly the same techniques as other gunsmiths of the time. His stocks were made by hand shaving and boring and his barrels were forged by hammers upon anvils and finished with rude drills and grindstones. The lock parts (see Figure 4.1) were ground and drilled, filed approximately to patterns, and fitted together. Whitney's innovation was to make the lock parts more uniform by the systematic use of hardened jigs, and to classify the work on a more intelligent and economical basis.

Assembling the lock parts was considered a crucial test of interchangeability. Because they could not be filed or milled after hardening, lock parts were traditionally assembled and fitted while soft, then marked or kept separate to avoid mixing after hardening. In order to be assembled after hardening, lock parts had to be made interchangeable.

Whitney systematized the work of firearms manufacture by making the parts in lots of large numbers and employing unskilled labor to file them, using hardened jigs to constrain their shape. Operations in his factory are described by Wilma Pitchford Hays.

> The several parts of the musket were, under this system, carried along through the various stages of manufacture, in lots of some hundreds or thousands of each. In their various stages of progress, they were made to undergo successive operations by machinery, which not only vastly abridged the labor, but at the same time so fixed and determined their form and dimensions, as to make comparatively little skill necessary in manual operations. Such were the construction and arrangement of this machinery, that it could be worked by persons of little or no experience, and yet it performed the work with so much precision, that when, in the later stages of the process, the several parts of the musket came

Summary of Epoch		American System
Size Trends	Introduced at Beretta (world)	1860 (1830)
	# of People (Min. Scale)	150
	Number of Machines	50
	Productivity Increase (over previous epoch)	3:1
	Number of Products	3
Nature of work	Standards for Work	Relative product
	Work Ethos	"Satisfice"
	Worker Skills Required	Repetitive
	Control of Work	Tight supervision of work
	Organizational Change	Staff-line separation
	Staff/Line Ratio	20:130
	Line Workers per Machine	3
Technology Keys	Process Focus	Precision: Repeatability (of machines)
	Focus of Control	Product conformance
	Instrument of Control	Go/No-Go gauges
	Rework (as fraction of total work)	.5

Table 4.1 Effects of the American System at Beretta

to be put together, they were readily adapted to each other, as if each had been made for its respective fellow. A lot of these parts passed through the hands of several different workmen successively, (and in some cases several times returned, at intervals more or less remote, to the hands of the same workman,) each performing upon them every time some single and simple operation, by machinery or by hand, until they were completed. Thus, Mr. Whitney reduced a complex business, embracing many ramifications, almost to a mere succession of simple processes, and was thereby enabled to make a division of labor among his workmen, on a principle which was not only more extensive, but also altogether more philosophical than that pursued in the English method. In England, the labor of making a musket was divided by making the different workmen the manufacturers of different limbs, while in Mr. Whitney's system the work was divided with reference to

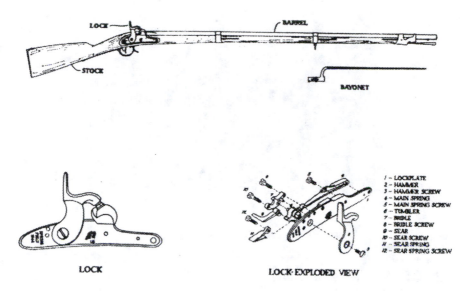

Fig. 4.1 Lock parts for 1842 musket [32, p 85]. Drawing by Steve Foutz

its nature, and several workmen performed different operations on the same limb.

It will be readily seen that under such an arrangement any person of ordinary capacity would soon acquire sufficient dexterity to perform a branch of the work. Indeed, so easy did Mr. Whitney find it to instruct new and inexperienced workmen, that he uniformly preferred to do so, rather than to attempt to combat the prejudices of those who had learned the business under a different system. [14, pp 53–54]

As a means to ensure precision in barrel manufacture, Whitney introduced "go" and "no go" gauges (Figure 4.2). The smaller of the two plugs was to fit into the barrel. If it did not, or if the large plug did fit into it, the barrel was rejected. Imposition of explicit standards improved the quality of arms, and in 1823 the Ordnance Department began requiring the use of go/no go gauges for arms inspection. [12, p 174]

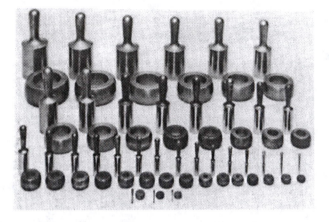

Fig. 4.2 A set of Go/No Go gauges [9]

4.2. Of Machines and Men

Of this period, Charles Fitch wrote that

> So far as machinery had been introduced, its construction was rude, and its use exceptional. Hand-shaving and chiseling for the stocks, and hand-forging, grinding, and hand-filing for the metal parts, constituted nearly all of the work.
>
> Apart from all consideration of the earliest usage of specific machines, it must be said that their introduction did not make itself felt as a great industrial agency until within twenty-five years past, in instance of which it may be stated that in 1839, there were at the Springfield armory about six men to one machine, and the ratio at other works seems to have been equally large; for of the private armories most reputed for early improvements one is stated at this time to have had but a single milling-machine, and that a rude one; and at another armory a single gang-saw profiling-machine was the principal stocking machine in use. It was some fifteen years later before the manufacture of milling, edging, and other important gun machinery was conducted on a scale sufficiently extensive for the general outfitting of large armories. [11, p 7]

The use of this machinery coupled with the use of water power to drive it had combined, as we saw in the earlier description of the Whitney factory, to reduce the skill requirements, though not necessarily the cost, of labor. Fitch observed that

> Relative to the skill required in the manufacture (of guns), since most of the work is special and done by the piece, few of the operatives may, in any case, be placed under the schedule caption of ordinary laborers. The foremen upon the several jobs or sub-contracts (who may be usually rated at 1 foreman to 30 or 40 operatives), the blacksmiths and the machinists proper, the tool-makers and the barrel straighteners, are considered skilled work-men, but the machine-tenders and other operatives, however pro-ficient in their special duties, are not so considered. The skilled men thus specified will generally constitute less than 20 percent of all. But in many factories much of the machinery is tended by experienced men, drawing the wages of skilled workmen, and the employment of unskilled labor, often adduced as an advantage due to improved machinery and the interchangeable system, seems largely available only on heavy contracts, when it may be utilized with a careful system of oversight. Machinery may contract the province of certain skilled trades ... but the ... increased fineness and accuracy required in the manufacture of fire-arms demands the most skillful and experienced oversight, and unskilled labor can only be employed with the best results upon limited portions of the work. Thus we find that at most of the larger armories the greater proportion of the operatives draw the wages of skilled men. [11, p 8]

The system lent itself to piece work and we find that many arms manufacturers subcontracted much of their work, either bringing con-tractors into their plants to work under local supervision or sending the work out to smaller shops.

4.3. The Colt Armory

Thanks to an instructive visit provided by Haven and Belden, we can see how the various aspects of the interchangeable system came together in the arms factory established in Hartford by Samuel Colt.[2]

> [The new armory] was finished and operations commenced in it in the Fall of 1855. As will be observed by the diagram, the ground plan of the principal buildings form the letter H. [See Figure 4.3] ...
>
> The motive power is located about in the center of the main building. It consists of a steam engine – cylinder, 36 inches in diameter, 7 foot stroke, fly-wheel 30 feet in diameter, weighing 7 tons ... which is rated at 250 horse power ... The steam is furnished from two cylindrical boilers, each 22 feet long and 7 feet in diameter. The power is carried to the attic by a belt working on the fly-wheel; this belt is 118 feet long by 22 inches wide, and travels at the rate of 2,500 feet per minute. [2]
>
> We now follow them to the armory proper, which, in the first place, is the second story of the front parallel. This is probably not only the most spacious, but the best arranged and fitted workshop extant ... On first entering this immense room, from the office, the *tout ensemble* is really grand and imposing, and the beholder is readily impressed with an exalted opinion of the vast mechanical resources of the corporation. The room is 500 feet long by 60 feet wide, and 16 feet high. It is lighted, on all sides, by 110 windows that reach nearly from floor to ceiling; it is warmed by steam from the boilers – the pipes being under the benches, running completely around the sides and ends; there are the perfect arrangements for ventilation, and sufficient gas burners to illuminate the whole for night-work. Running along through the center is a row of cast-iron columns, sixty in number, to which is attached the shafting – which here is arranged as a continuous pulley – for

---

[2] The following is primarily from [13, pp 352-358]. It overlaps substantially with [2], and several paragraphs are from that source where indicated.

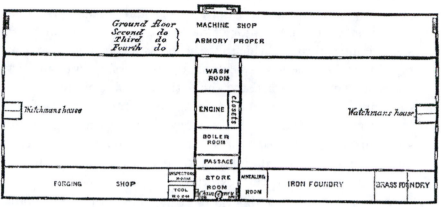

GROUND PLAN OF THE ARMORY.

Fig. 4.3 Floor plan of Colt armory, 1855 [13, p 352]

driving the machines, as close together as possible, only allowing sufficient space to get around and work them. [2]

The whole of this immense floor space is covered with machine tools. Each portion of the fire-arm has its particular section. As we enter the door the first group of machines appears to be exclusively employed in chambering cylinders; the next is turning and shaping them; here another is boring barrels; another group is milling the lock-frames; still another is drilling them; beyond are a score of machines boring and screw-cutting the nipples, and next to them a number of others are making screws; here are rifling machines, and there the machines for boring rifle-barrels ... Nearly 400 [machines] are in use in the several departments.

It is unnecessary to describe all the operations performed by the machines; a few will render the whole understandable. Taking the lock-frame, for instance; they commence by fixing the center, and drilling and tapping the base for receiving the arbor or breech-pin, which has been previously prepared – the helical groove cut in it, and the lower end screwed – once grasped is firmly fixed into position, furnishing a definite point from which all the operations are performed, and to which all the parts bear relation. The facing and hollowing of the recoil shield and frame, the cutting and sinking the central recesses, the cutting out all the grooves and orifices, planing the several flat surfaces and shaping the curved parts prepare the frames for being introduced between hard and steel clamps, through which all the holes are drilled, bored and tapped for the various screws; so that, after passing through thirty-three distinct operations, and the little hand finishing required in removing the burr from the edges, the lock-frame is ready for the inspector. The rotating, chambered cylinder is turned out of cast-steel bars, manufactured expressly for the purpose. The machines, after getting them the desired length, drill center holes, square up ends, turn for ratchet, turn exterior, smooth and polish, engrave, bore chambers, drill partitions, tap for nipples, cut pins in hammer-rest and ratchet, and screw in nipples. In all there are thirty-six separate operations before the cylinder is ready to follow the lock-frame to the inspector. The barrel goes through forty-five separate operations on the machines. The other parts are subject to about the following number: lever, 27; rammer, 19; hammer, 28; hand, 20; trigger, 21; bolt, 21; key, 18; sear spring, 12; fourteen screws, seven each, 98; six cones, eight each, 48; guard, 18; handle-strap, 5; stock, 5. Thus it will be observed that the greater part of the labor is completed in this department. Even all the various parts of the lock are made by machinery, each having its relative initial point to work from, and on the correctness of which the perfection depends.

[The upper floor] is designated the Inspecting and Assembling Department. Here the different parts are most minutely inspected; this embraces a series of operations which in the aggregate amount to considerable; the tools to inspect a cylinder, for example, are fifteen in number, each of which must gauge to a hair [see Figure 4.4]; the greatest nicety is observed, and it is absolutely impossible to get a slighted piece of work beyond this point. On finishing his examination, the inspector punches his initial letter on the piece inspected, thus pledging his reputation on its quality.

On their final completion, all the parts are delivered to the general store-keeper's department, a room 60 feet wide by 190 feet long, situated in the second story of the central building, and extending over the rear parallel. All the hand-tools and materials (except more bulky kinds) are distributed to the workmen from this place; several clerks are required to parcel the goods out and keep the accounts; in fact, it is a store, in the largest sense of the term, and rather on the wholesale principle at that. On the reception of finished, full sets of the parts of the pistols, they are once more carried up to the assembling room; but this time to another corps of artisans. Guided by the numbers, they are once more assembled.

We have followed ... through about 460 separate processes of manufacture, which, in the usual course pursued would have occupied from three to four weeks of time.

During the time of our visit we were informed that scarcely less than one hundred thousand weapons were at that moment in the various stages of progress, yet the whole number of employees was little less than six hundred who, by the aid of mechanical contrivances, turn out an average of two hundred and fifty finished arms per diem.

In rough numbers it might be stated that supposing the cost of an arm to be 100; of this the wages of those who attended to and passed the pieces through the machines was 10 per cent, and those of the best class workmen engaged in assembling or putting together, finishing and ornamenting the weapons was also 10 per cent, thus leaving 80 per cent for the duty done by the machinery.

A majority of the machinery was not only invented, but constructed on the premises. When this department was commenced, it was the intention of the Company to manufacture solely for their own use. Some months since, applications were made by several foreign Governments to be supplied with machines and the right to operate them. After mature deliberation, it was concluded to supply the orders, and on the day of our visit we saw a complete set of machinery for manufacturing fire-arms, that will shortly be shipped to a distant land. The Company have now determined to incorporate this manufacture as a branch of their regular business.

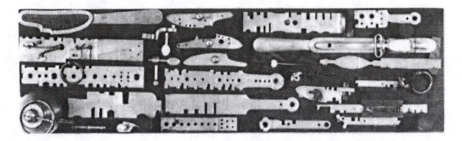

Fig. 4.4 Set of Inspection Gauges for US Rifle Model 1841 [5]

In the American arms factories, as exemplified by the Colt Armory, the foremen were contractors who hired their own help as subcontractors to produce the various parts of the gun. When a man had made his contract, he was provided with a machine and left on his own to complete the order. Many of the improvements in metal working methods, most of them undocumented, derived from the zeal of individuals who applied their ingenuity to the machine in front of them in order to realize the savings that would result from increased productivity. The British Royal Commission on the American System reported that "in the adaptation of special apparatus to a single operation in almost all branches of industry, the Americans display an amount of ingenuity combined with undaunted energy." [26, p xii] These improvements were seldom patented. Most became common knowledge, and were appropriated by others who carried the improvements still further.

With the emphasis of manufacturing during this period on interchangeability of parts, the focus of control shifted from product *functionality* to product *conformance*. Though still patterned after a model, a piece was expected to conform not just to the pieces it was to mate with in a given rifle, but to those same pieces in every gun of a given design. Accuracy in this system, which might be as close as a thirty-second or sixty-fourth of an inch, was ensured by an elaborate system of patterns, guides, templates, gauges, and filing jigs (Figure 4.4).

Writing in 1880 on the degree of uniformity then being achieved Fitch observed that

If gun parts were then called uniform, it must be recollected that the present generation stands upon a plane of mechanical intelligence so much higher, and with facilities for observation so much more extensive than existed in those times, that the very language of expression is changed. Uniformity in gun-work was then, as now, a comparative term; but then it meant within a thirty-second of an inch or more, where now it means within half a thousandth of an inch. Then interchangeability may have signified a great deal of filing and fitting, and an uneven joint when fitted, where now it signifies slipping in a piece, turning a screw-driver, and having a close, even fit. [11, p 4]

As product conformance relied on repetition, the focus of process control became *repeatability*, that is to say, each execution of a process was expected to produce the same part with high precision. Of course, actual precision was not always sufficient, and quality was achieved through 100% inspection for functionality: "[In the prover's department] each chamber is loaded with the largest charge possible, and practically tested by firing ..." [2]

4.4. The American System Abroad

Robbins and Lawrence, a machine tool builder that had perfected the use of the American System, seeing the commercial potential of their work, exhibited their guns at the Crystal Palace Industrial Exposition in 1851, where the rifles garnered an award and attracted such attention that Parliament was induced to send a commission to the United States to study "the American system" of interchangeable manufacture and secure the machinery necessary to introduce it at the Enfield Armory near London. The company received an immediate order for 20,000 Enfield rifles with interchangeable parts and 157 metal working machines to equip the armory at Enfield. [30, p 138] With this order, Robbins and Lawrence became the first large-scale exporter of machine tools.

When Samuel Colt set up his own integrated factory near London, the American System was placed on display and soon held in awe by all the major arsenals in Europe. Pratt and Whitney, one of the major machine tool builders that supplied the Colt factory, was soon receiving orders from almost every country in Europe for machinery (see Figure 4.5) with which to set up factories.

Giuseppe Beretta had seen the superiority of the new system of manufacture in the Prussian arms factories, which had acquired from Pratt and Whitney the entire manufacturing system, lock, stock and barrel. Not wanting to fall behind, Beretta, in 1860, had Pratt and Whitney build an integrated factory in Gardone. With this one stroke he had the largest arms factory in all of northern Italy. The two hundred workers in the Beretta factory were soon turning out eight thousand sporting guns and three thousand military rifles per year.[3]

In 1881 Beretta was awarded a medal for its innovations in the factory system. It was the only firm that took in iron and wood through one door and sent out finished arms through the other. The company sold not only in Italy, but throughout the world, particularly in the regions of the East. Its precision of manufacture was such that the company was able to offer a guarantee against any and all breakdowns and defects for one year. Beretta introduced a number of new products and watched its volume of manufacture grow.

The one drawback in all of this progress had to do with the nature of work. The activities at the workstation, as we saw earlier in the description of the Colt factory, did not require much skill. There were now two kinds of workers: those who built, maintained, set up, and improved machines; and those who turned out parts by the hundreds every day. Together, the separation of staff and line work, the specialization of line work, and the elimination of skill at the workstation had the effect of creating a competitive market for labor, which tended to depress wages. Gardone saw its first strike by workers, which interrupted work for several days, in 1878. The subsequent organization of unions and establishment of a labor cooperative considerably improved the tenor of workers' lives.

Work was now reduced to *managing machines and output*. Labor was a corporate entity that *executed procedures*; engineers conceived of the tasks. The separation of conception and execution of work was now complete. Mechanical work could now be abstracted, studied in

---

[3] Implying roughly 5 man-days per firearm.

Fig. 4.5 Milling machines for gun-making, circa 1885; note multiple pulleys to adjust speeds [17, p 142]

isolation from the plant, and then developed in the plant and reproduced by other workers. Mechanical work was becoming a science.

Although it was conceived as a system for the *manufacture of interchangeable parts*, the American System's major contribution to process control was the notion of *mechanization of work*. Whereas the English System saw in work the combination of skill in machinists and versatility in machines, the American System introduced to mechanisms the modern scientific principles of *reductionism* and *reproducibility*. It examined the processes involved in the manufacture of a product, broke them up into sequences of simple operations, and mechanized the simple operations by constraining the motions of a cutting tool with jigs and fixtures. Verification of performance through the use of simple gauges ensured reproducibility. Each operation could now be studied and optimized.

In the context of the American system it was necessary to attend not only to the construction of special purpose machines, but also to the interrelationships among them. In order to design and build a collection of special purpose machines for manufacturing a component, one had to conceive of an entire system of manufacture. This entailed being an architect of a collection of mechanisms, as well as bringing scientific principles to the study of mechanisms. Manufacturing was now front-end loaded, that is to say, the most significant aspects of cost and quality of components were established prior to the production of the first unit. The importance of special-purpose machinery to such a system cannot be overstated.

Over the next hundred years, these simple mechanisms would be elaborated, eventually becoming self-acting and capable of great precision and versatility. As understanding of the principles of mechanization became diffused with the increasing specialization of machines, variability of work returned once more to labor.

The next major intellectual watershed would be crossed when the new science of machinery was extended to human labor. Application of the principles of machine movement to human work yielded a new *scientific management* of work, the impact of which on the organization of work at the Watertown Arsenal we will now examine.

5

The Taylor System

To Frederick Taylor (1856–1915) falls the distinction of doing for work what a century of refinement had done for machinery. Taylor recognized that the machinery available at the end of the nineteenth century was capable of more than workers were getting out of it. Worker-related activities, he realized, were limiting the speed and efficiency of the machines. The idea that these human activities could be measured, analyzed, and controlled by techniques analogous to those that had been successfully applied to physical objects was the central theme in what Taylor was to put forth as a theory of "scientific management."

As conceived and practiced by Taylor, scientific management was concerned with industrial work, particularly the work of machine shops in metalworking establishments. Taylor was concerned almost exclusively with organization at the shop level, from the superintendent and foreman down.

Although he shared with Church, Halsey, and Towne in the United States, and Slater Lewis in Britain, an interest in incentive wage payments as a means of increasing productivity, Taylor took a different approach. He viewed work as an object and studied it as if it were a physical, mechanical entity. In the Taylor scheme of things, job times were determined not by past experience, but with a stopwatch.

Standard times were to be set for each job and a standard rate of output determined. This involved two elements: job analysis, and time study.

Job analysis consisted in breaking each job down into small elements and distinguishing those elements that were essential for the performance of work from those that were superfluous, or "waste." The waste was to be eliminated. Once the elements essential for the performance of work had been isolated, they were classified functionally in order that functional specialists could carry out different aspects of the job. For instance, a machinist assigned the task of turning down a piece of metal to certain dimensions on a lathe might find that his cutting tool needed sharpening. Taylor considered the skills associated with sharpening tools to be functionally different from the skills of a machinist and, consequently, separated them. Similarly, it was not part of the machinist's job to determine correct speed or the correct angle of a cut. Even more obviously, it was not part of his job to obtain materials or tools from the storeroom or to move work in progress from place to place in the shop or to do anything but turn the piece of metal on his lathe. Job analysis, as Taylor interpreted it, almost invariably implied a narrowing of the functions included in the job, an extension of the division of labor, and trimming off of all variant, non-repetitive tasks.

The second basic element in Taylor's system was time study. After a job had been analyzed into its constituent operations these were timed with a stopwatch. By adding the elementary times for each operation, a total time for the whole job was calculated. Operations were classified into two types: machine time, and handling time. Machine times could be precise, as they depended on physical characteristics of the metal being worked, the cutting instrument, and the machine tool. Handling times, which referred to the time taken by an operator to set up work on the machine and remove it after the work was completed, varied widely among operators.

Nor were the machines themselves neglected by the Taylor system. Inasmuch as the speed of operators was largely determined by the speed of the machines as driven from a central location by belts, pulleys, and shafts (see Figure 5.1), Taylor considered the standardization and control of these systems at their optimal level of efficiency essential. To this end he established the activities of belt maintenance and adjustment as a separate job and prescribed methods for scientifically determining correct belt tensions.

Taylor employed Henry Gantt and Carl Barth to assist with the specification of optimum cutting speeds. "Taylor succeeded in determining empirically," explains Aitken,

> by a prolonged series of experiments, the optimum relationship between all the variables that influenced the rate at which metal could be cut on a lathe: the depth of cut, feed, speed, and type of tool, the hardness of the metal, the power applied to the machine, and so on. These results were plotted on graph paper, giving a set of geometric curves from which the proper speed of the lathe could be determined when the values of all other variables were known. This method of solving the problem was, however, too slow and inconvenient for ordinary shop use. Barth ... reduced the relationships discovered by Taylor and Gantt to a mathematical equation and transferred the functional relationships involved to specially made slide rules, which made it possible to determine the correct speed of a machine tool quickly and with all the accuracy required for practical use. [1, p 33]

Together with job analysis and time study, this systematization of machinery introduced a level of precision previously unknown. With the application of these concepts, work could truly be said to be standardized.

Because company records for Beretta are sketchy for this period, to explain the profound implications of the Taylor System, not only for the firearms industry, but for all of manufacturing, we must turn to changes in manufacturing practice at the Watertown Arsenal as described by Aitken. Of Beretta, we know that following the First World War Pietro Beretta completely renovated its factory, introduced machinery compatible with the innovation of high-speed tool steel, and incorporated the principles of scientific management as enunciated by Fayol and Taylor. We know little about how Taylorism was adopted by Beretta. Indeed, beyond such aggregate statistics as the plant's

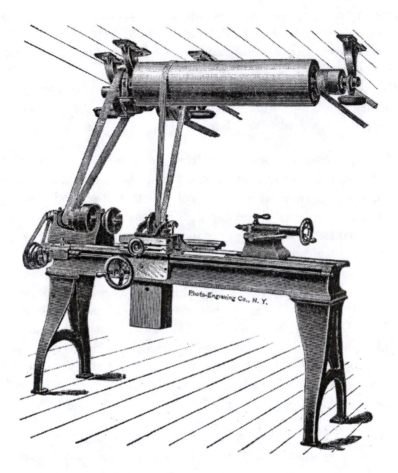

Fig. 5.1 Belt-driven lathe, circa 1885 [1, p 68]

tripling in size and realization of a tenfold increase in production over a period of fifteen years, we know little of Beretta's progress during the Taylor era. What we know is summarized in Table 5.1.

5.1. Taylorism at the Watertown Arsenal

At the beginning of the twentieth century, the Watertown Arsenal employed some two hundred and fifty people in its machine shops and

Summary of Epoch		Taylor System
	Introduced at Beretta (world)	1928 (1900)
	# of People (Min. Scale)	300
Size Trends	Number of Machines	150
	Productivity Increase (over previous epoch)	3:1
	Number of Products	10
	Standards for Work	Work standards
	Work Ethos	"Reproduce"
	Worker Skills Required	Repetitive
Nature of work	Control of Work	Loose of work/ tight of contingencies
	Organizational Change	Functional specialization
	Staff/Line Ratio	60:240
	Line Workers per Machine	1.6
	Process Focus	Precision: Reproducibility (of processes)
Technology Keys	Focus of Control	Process conformance
	Instrument of Control	Stop watch
	Rework (as fraction of total work)	.25

Table 5.1 Effects of the Taylor System at Beretta

another fifty in its foundry. The general condition of the arsenal at that time is described by Aitken.

> Forty percent of the machine tools had been in service for fifteen years or more, and many of them for over twenty years. All had been designed when carbon tool steel was the only type of tool steel available, and the system of belts and pulleys had been set up on that basis. Throughout the plant there was a serious shortage of small parts, such as belts, straps, and clamps. Handling facilities, such as cranes and trolleys, were inadequate ... [In the foundry] the only mechanical assistance available to the molders were two cranes and a few pneumatic rammers. For the rest, the work was entirely by hand.
>
> There was a headquarters office staffed by clerks, whose principal function was correspondence, and an engineering division,

an innovation introduced ... in 1908, which was responsible for the making of cost estimates, the preparing of blueprints, and similar tasks ... In the day-to-day operations of the arsenal the master mechanic and foreman were largely left to their own devices to allocate jobs and get out the work.

Apart from its organizational defects, the machine shop at Watertown was technically far from up-to-date in 1909. Into this machine shop there was introduced a major innovation, high-speed tool steel, probably the most revolutionary change in machine shop practice within the memory of anyone living at the time.

High-speed steel was no minor change which could be introduced in one section of the arsenal and then forgotten. The whole arsenal had to be geared to the pace which it set. [Colonel] Wheeler took a lathe which, using one of the old-style carbon steel tools, could be made to remove a maximum of two hundred pounds of metal from a casting in an eight-hour working day. He put a high-speed tool in the lathe and set it to work on the same job under the same conditions, except that the speed, feed, and depth of cut were altered to suit the new tool. The lathe removed precisely ten times as much metal – two thousand pounds – in the same period of time. This was probably an exceptional case, for the usual increase seems to have been 200 or 300 per cent.[1]

It was possible for a machine shop to purchase a stock of high-speed tools, use them in its metal-cutting operations, and yet continue to turn out work at much the same rate as before ... A shop which did this would find the results disappointing; the only obvious change would be that the tools would not need to be reground so frequently. To purchase high-speed tools was one thing; to use them correctly, so that their full potentialities could be realized, was another.

A machine shop which adopted high-speed steel and knew what the new steel could do was faced with the necessity of a

---

[1] Conventional steel loses its sharp edge at temperatures around 400°F (200°C), and therefore a slow cutting speed (often in conjunction with lubrication) must be used to keep the temperature always below this. High-speed steel, a steel alloy with less than .5% carbon, maintains hardness at higher temperatures, with the implications discussed in the text. Taylor's development of this steel is discussed further in [7].

complete reorganization. First, few of the machine tools built to use the old steels could be run at the pace and with the power which the new steel demanded. Hence the necessity for rebuilding and redesigning machine tools, systematizing belt maintenance and repair, and increasing power capacity. Secondly – a considerably more intractable problem – few of the machinists and foremen who had grown up in the carbon steel era had any conception of what the new steels could do. Hence the necessity for Barth's slide rules and the prescribing by management of speeds and feeds which, to men of the older generation, were literally fantastic. And third, since the use of high-speed steel meant very large reductions in machining times, handling times (the time taken to set up a job in the machine before machining and to remove it afterwards) came to represent a higher proportion of total job times. [1, pp 86-87, 102-103]

From the rising relative importance of labor time came the necessity to examine the labor component of work, specifically the uncertainty introduced by labor. System variance was due not so much to the accuracy and precision of machine tools as to labor's ability to reproduce a given procedure. Fortunately, Taylor paid as much attention to the procedures of manufacture as to the processes.

After consulting with Colonel Wheeler, the superintendent of the arsenal, Carl Barth, a protégé of Taylor, undertook a "complete reorganization of the whole shop management along the lines of the Taylor System," an endeavor that "fell into four parts or phases:

(1) Reorganization of the storeroom and toolroom,

(2) Creation of a planning room and establishment of a routing system,

(3) Re-speeding and standardization of machine tools, and

(4) Installation of an incentive wage system based on time study and task setting."

Aitken elaborates:

The first point of attack was the storeroom ... A new system for accounting for stores issued was instituted ... to provide automatic checks against excessive or duplicate issues of materials. Barth also recommended the installation of the "double bin" system, whereby two separate but adjoining bins ... were used for each article in store, one the receiving and the other the issuing bin. When all of the articles in the issuing bin had been distributed, it became the receiving bin and what had been the receiving bin became the issuing bin. As the bins were successively emptied, the tags showing all issues from them were sent to the property division, where they were checked against the stock sheets. This simple but highly effective system provided an automatic inventory of stores: the quantity of an article on hand was verified each time the issuing bin for that article was emptied. The tool room also was reorganized. ... The toolmaking section was separated from the tool-issue section and a foreman appointed to supervise the manufacture and care of tools. ... Orders [were] placed for the standard Taylor tool-forging and -grinding equipment. A special allotment [was made] ... for the installation of tool-managing facilities ... [This] was supplemented ... by a further allotment for the purchase of high speed tools.

An important series of changes made during the same period was the establishment of standard procedures for the inspection and maintenance of the belting which drove the machine tools. [Barth] recommended the purchase ... of a belt bench, a set of ... belt scales, and a wirelacing machine ... At the same time a special belt-maintenance gang was formed, and its foreman ... sent for instruction ... A great deal of the old belting was replaced with new and in some cases heavier belting. This made it possible to run machines at higher speeds and with greater power, so that full advantage could be taken of the cutting powers of high-speed steel, and also prepared the way for Barth's later standardization of cutting speeds and feeds. By the end of April 1910 the belt-maintenance system was in full operation and belt failures during working hours had been practically eliminated.

It was decided that two ... gun carriages should be the first products to be put through the machine shop "under the Taylor system." This meant ... that their manufacture would be routed from the planning room, since Barth had not yet begun his work on the machine shop and no time studies had been made. The two six-inch carriages were to serve as pilot projects, principally to give the planning room staff and the machinists their first taste of centralized routing.

A considerable amount of preparatory work was necessary. Assembly charts were drawn up, containing a detailed analysis of the operations required to produce each component that went into a complete gun carriage. On the basis of these charts the planning room decided upon the sequence of operations which each component was to follow, the dates at which each operation should be started and completed, and the order in which each component should be moved from workplace to workplace, up to final assembly. These individual analyses were then brought together on a single route sheet, which formed the master timetable for the whole project ... The route sheet was then turned over to clerks, who prepared the individual job cards, move tickets, and so on which would be required for the execution of the work. These cards, together with the master route sheet, were then passed to the route-sheet clerk, who filed them ready for use when required.

By the middle of April 1910 practically all of the work in the machine shop was being routed from the planning room.

While the planning room was getting into action, Barth took up the ... rehabilitation of machine tools. This involved four principal phases: the diagramming of the individual machine tools, the rebuilding and redesigning of obsolete or unsuitable tools, the standardization of ancillary equipment, and the prescribing of speeds and feeds. ... An extensive series of tests on different types of high speed-steel was conducted and, during a slack period of work early in 1910, the opportunity was taken to relocate a number of the machine tools, to bring together in one section of the shop all equipment of the same type. Several of the larger machine tools were provided with individual electric motors, while the smaller ones were arranged so that they could be driven in groups.

Diagrams were prepared for all machines in the machine shop, showing the driving arrangement, feed gears, and so on. This process necessitated the measuring of all the machines, their gears and pulleys, and the study of the diagrams to insure uniformity and the proper relationships in their speeds and feeds. A considerable amount of redesigning and rebuilding was done, particularly of cone pulleys and gears, first on the lathes and then on the drills, planers, shapers, slotters, and other machines.

At the same time arrangements were made to standardize all the ancillary equipment used on the machine tools. The sockets for boring bars were standardized so that all boring bars would fit all sockets, and the slots in the faceplates of the lathes, planers, and other tools were cut to a single size. All the tee bolts were made to fit this standard slot, to avoid any delay in finding the right bolt for a particular job. To achieve uniformity in cutting tools, *the workmen were forbidden to grind their own tools ...* Instead, the toolmaking department was to forge and grind all tools to the standard Taylor specifications and by the use of Taylor-designed equipment. The tool posts on the lathes were altered and strengthened so that they could be used with these tools. [emphasis added]

From August 1910 until ... June 1912, Barth spent four days each month at Watertown. These monthly visits were entirely taken up by work on the preparation of slide rules for the machine tools and in instructing the planning room in their use.

[Dwight] Merrick was hired to carry out time studies, by means of a stop watch, of the various jobs in the machine shop and to teach time-study work to certain members of the planning room staff. These time studies had three chief purposes: (1) to simplify work by the elimination of superfluous motions, (2) to set a standard time in which each job ought to be done, and (3) to provide the basis for a payments scheme which would furnish an incentive for the workmen to do the job in the standard time. Merrick's work was essentially an extension of the improvement and standardization work which Barth had begun. What Barth had done for the machines, Merrick was now to do for the men.

Merrick's task was considerably more complicated than anything Barth had undertaken. Barth could examine a machine tool, change its gears and its belting, and reset it to run at a higher speed in complete confidence that the machine would do what he wanted it to. But Merrick had to take some account of the fact that the men would not work at the pace he prescribed unless they wished to do so. He had to face the problem of motivation.

The answer which the Taylor system provided to this problem was an incentive-payments scheme. If you wanted men to work at a certain pace, you promised them a financial reward if they did so; the problem was no more complicated than that.

Taylor doctrine did not entirely overlook the possibility that the introduction of time study in a plant might cause trouble. The stop watch had not yet become a symbol of all that was detestable to organized labor in the Taylor system, but it was already realized among the Taylor disciples that the purposes of time study could easily be "misunderstood" ... It was considered vital that no time studies be attempted until all working conditions had been brought up to a high level of efficiency and standardized at that level. There were two reasons for this. First, if conditions were not standardized, then the job itself was not standardized and could not be scientifically timed. It would serve no purpose to set a time on a job today if tomorrow the machine might be running at a different speed, or if the workman had to wait around at the window of the storeroom until his material was ready, or if he had to leave his machine idle while he ground his tools. And secondly, it was believed that there would be less resistance to time study if the men being timed had grown accustomed to seeing a whole series of changes being made in their working conditions, all of which made their work easier.

Unlike other parts of the Taylor system which, as soon as they were installed, affected the organization and operation of the entire shop, time study was introduced gradually and almost imperceptibly.

Merrick continued in regular employment at Watertown until June 1913, by which date several of the arsenal's employees had been trained to take his place as time-study experts. By the end of June 1913 ... the Taylor system at the arsenal was on a self-sustainingbasis ... It was the arsenal's regular system, and all that remained was to extend and complete it. [1, various pages. See also 7]

In 1900, Frederick Taylor and Maunsel White demonstrated their high-speed, chromium–tungsten steel at the Paris Exposition. [25, p 297]. Though it ran red hot, the metal did not soften or dull. As the structure of the machines at that time could not withstand the stresses induced by the forces of running so hot, heavier machines, five times as powerful, were built to exploit this innovation. Taylor and White took Europe by storm and soon Taylor's notions of scientific management were well publicized. The ascendancy of the American System of Manufacture and its culmination in the Taylor System were now well known. By the turn of the century, the leading engineering journals were full of the new gospel and supporting their preachments with examples of successful innovation.

5.2. Beretta's Belated Adoption of Taylorism

Taylorism arrived late at Beretta. With the death of Giuseppe in 1903, Pietro Beretta became head of the firm, a position he held until his own death in 1957. The latter Beretta greatly enlarged the factory, equipped it with the most modern equipment, and brought out many innovative products. Under his direction, the firm rose to international prominence in less than three decades.

Beretta installed three hydroelectric plants on the Mella River, the first in 1908 and the last in 1939, which contribute to the firm's energy supply to this day. With the availability, after about 1910, of machine tools capable of using high-speed steel, Beretta gradually began to renovate the equipment in his plant (see Figure 5.2).

Then a sharp decline in demand for sporting rifles in the years preceding World War I created a crisis among the manufacturers in the Trompia Valley. In the lull before the storm Beretta and his chief

Fig. 5.2 The Beretta factory, circa 1939 [18, p 230]

engineer, Tullio Marengoni, threw themselves actively into the development of new designs. A patent entitled "Innovations for Automatic Pistols," obtained by Beretta in 1915, was the conceptual cradle for a family of weapons that would make the company renowned the world over.

Although with the war, production of military arms shot up, there was no physical expansion of the facility. Some modern machines were added, but Taylorism did not take hold at Beretta until the late 1920s. From 1919 until 1922 life in Italy was torn and threatened by all manner of violence. Beretta was one of the few factories in Brescia that was not taken over by rebellious workers during this time.

Rationalization of production was the talk of Europe in the 1920s. Although the teachings of Taylor and Fayol found a receptive ear in Pietro Beretta, Taylorism was not adopted at the Beretta factory until 1928, when machinery acquired from the firm Fabbrica d'Armi Lario (called FALC), in Camerlate, near Como, was transferred to Gardone, necessitating a reorganization of the plant and layout. The size of the firm subsequently doubled and with the increased scale came the functional specialization of Taylorism.

The Ethiopian war of 1935 and Italian intervention in Spain the year after brought a flood of large orders for Beretta's highly successful automatic pistol, the Model 34. Taylorism and mass production of the Model 34 fit well together and the new organizational form took hold at Beretta.

It is difficult to say whether Beretta's rapid growth resulted from the implementation of Taylorism or from the requirements for mass production of the Model 34, which led naturally to the introduction of Taylorism. Indeed, it may be that there is no connection at all. But occur together they did and the result was a dramatic increase in labor productivity. Methods engineering was introduced and, though the size of the work force only doubled, the size of the staff group increased three-fold. Most of the impact of methods engineering was in the reduction of rework, which dropped from 40% to 25%.[2]

At Beretta, as elsewhere, Taylorism fundamentally altered the ethos of work. In the Beretta factory separation of line and staff was complete; the stop watch, the efficiency expert, and productivity measures were there to stay. Though belatedly, Taylorism had met Beretta and now they were one.

Let us now examine the changes wrought by the Taylor system on the workstation and process control at the workstation.

(1) The **scope of work** was restricted under the Taylor system. Material handling, tool sharpening, belt tightening, oiling machinery, work set-up, and other secondary activities were no longer done by the operator, but by specially trained people, each of whom performed only a single function. Such functional specialization was not entirely new, but it was greatly expanded by Taylor. It extended to secondary opera-

---

[2] Forty percent rework means that 40% of total work time was devoted to rework, not that 40% of the parts were rejected.

tions a concept begun with the American System of Manufacture.

(2) **Worker discretion** was eliminated. The Barth slide rule specified exactly how an operation was to be performed and tasks were set by an "efficiency expert." Work was broken down into small parts and standardized. There was "one best way" to perform a task, which was specified by an outside observer, and the worker's sole responsibility was to execute the procedure that constituted that one best way. System variance was reduced by eliminating worker discretion and making the task reproducible. This was Taylor's lasting contribution.

(3) **Control of work** was now in the hands of management, which could compare the quantity of work performed against a predetermined standard and monitor worker effort. Much has been written on the aspects of management control of work and the setting of standards. Senate testimony on Taylorism at the Watertown Arsenal documents the resulting antagonism between management and labor. [15]

With Taylorism we had crossed another intellectual watershed – the study of procedure of manufacture independently of the process of manufacture. How one set up a tool on a machine was largely independent of the process, hence tool set-ups could be studied in their own right. The efficient layout of a plant was now independent of what was made. Such separations gave rise to the field of industrial engineering, which was quite distinct from its parent, mechanical engineering.

In disassociating a product's function from the process used to make it, the American System had effectively endowed the manufacturing process with a life of its own. Taylor's further separation of procedure from process reflected his recognition that the chief constraint on the speed and efficiency of a machine was its human operator. The underlying logic of scientific management as practiced and promoted by Taylor – consistent control achieved by scientifically modeling the steady state behavior of workplace processes – yielded hitherto undreamed of levels of precision by eliminating waste in execution and reducing variance in performance.

But in applying scientific method to the study of steady state conditions, Taylor overlooked the statistical nature of the world. Lacking was a scientific understanding of departures from steady state. Nearly a half century would pass before recognition of the many sources of variation in a large factory – including imprecision of machines, scale effects, and environmental factors – would give rise to the technique that would fill this void, namely, statistical process control (SPC).

Taylor's principles were well suited to the largely static factories of the turn of the century. But in the dynamic world spawned by more frequent innovation in, and increasing automation of, machine tools, the "one best way" to execute a process might change month to month, week to week, even day to day.

6

The Statistical Process Control Era

The first three epochs emphasized increasing mechanization in a world that was, at least ideally, static – doing the same tasks again and again, as efficiently as possible, at increasingly high volume. Discretion was progressively removed from workers, and knowledge about their tasks was subdivided and given to specialists, removing it from the shop floor. Minimum efficient scale increased from a handful in the craft era, to 40 people in the English System, to 300 in the Taylor era, while output per person grew by a factor of 40. In contrast, in the last three epochs, while the tools continued to become more mechanized, knowledge about the work was returned to workers and their discretion increased. The key goal shifted from efficiency at high volume to coping with a dynamic world of rapid changes such as high product variety and rapid product introduction.

Packed into the little more than half a century following the First World War are the three eras of the dynamic world: Statistical Process Control (SPC), the Numerical Control era, and the age of Computer-Integrated Manufacturing. As Beretta was by this time a leader in the technology of arms manufacture, the discussions of these eras focus specifically on the Beretta factory.

In the 1950s, following the establishment of NATO, Beretta was licensed to manufacture the semi-automatic Garand M1 rifle. A decade earlier, the Garand had transformed the manufacturing system at the Springfield Armory. It would do the same at Beretta, but with a fundamental difference.

Most of the equipment at Springfield when it received the contract to manufacture Garands was of World War I vintage. Because the armory had lagged behind in the adoption of the new technology, which emphasized the integration of multiple operations in a single machine, a massive program of equipment renovation and plant modernization had to be undertaken. (See Figure 6.1 for photos of the machinery used to manufacture the Garand at Springfield and Figure 6.2 for an illustration of the machining operations required on a particular piece.)

Beretta's contract, coming a decade later after the Second World War, had an impact that went beyond the renovation of equipment. The M1 required tighter tolerances by an order of magnitude than any the company had previously achieved, and the specified degree of interchangeability of components was 100%. Machines for producing the Garand therefore had to exhibit both accuracy (for interchangeability) and precision (for tight tolerances). Beretta chose to build its own machines on principles established by Garand. As Beretta was not a machine tool builder, this experience exerted a profound influence on both the company's manufacturing system and its organization of production. The effects of SPC at Beretta are summarized in Table 6.1 .

6.1. Monitoring Variation and Its Sources

Because the machines that Beretta built did not have the process capability of, for instance, the Kingsbury multi-station machines for drilling and reaming, the machining process had to be constantly monitored for any deviation from the prescribed process settings. Machines with a tendency to "wander" demanded a new technique to enhance process capability. Thus it was that statistical process control, invented in the United States in the 1930s in the electrical equipment industry, was adopted by Beretta before it was used at the Springfield Armory.

Fig. 6.1 Springfield Arsenal machinery for Garand M1 rifle [4, pp 8, 11]

Though it seemed, at the time of its introduction, an innocuous enough change, statistical process control radically altered the organization of work at Beretta. Meant to ensure process stability, SPC required only that process behavior for a sample of parts be recorded on charts at specified intervals of time. Yet we shall see how, over a period of only five years, it changed the ethos of manufacturing management at Beretta and with it the organization of work.

It is helpful, at this point, to clarify the concepts of process capability and process stability. The precision of which a machine is capable is defined by the standard deviation of the random variation in its performance. It is a property of the machine. The tolerance required

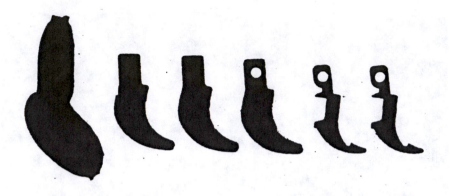

Fig. 6.2 Sequence of machining operations for M1 safety [3, p 121]

	Summary of Epoch	Statistical Process Control
Size Trends	Introduced at Beretta (world)	1950 (1930)
	# of People (Min. Scale)	300
	Number of Machines	150
	Productivity Increase (over previous epoch)	3:2
	Number of Products	15
Nature of work	Standards for Work	Process standards
	Work Ethos	"Monitor"
	Worker Skills Required	Diagnostic
	Control of Work	Loose supervision of contingencies
	Organizational Change	Problem-solving teams
	Staff/Line Ratio	100:200
	Line Workers per Machine	1.3
Technology Keys	Process Focus	Precision: Stability (over time)
	Focus of Control	Process capability
	Instrument of Control	Control chart
	Rework (as fraction of total work)	.08

Table 6.1 Effects of SPC at Beretta

by a workpiece might be more or less than a machine is capable of achieving. *Process capability* is the relationship between the precision of the machine and the tolerance required by a workpiece. A common measure of process capability is defined by the ratio:

Cp = tolerance/(6 * the standard deviation).

Process stability refers to the frequency with which a process goes out of control. It is a measure of machine reliability, and is unrelated to process capability. A machine might be very capable but unreliable, or vice versa.

Tolerance requirements for a variety of operations on the M1 are .001 inches, as can be seen in the operations sheet (Figure 6.3). Finish reaming and rifling operations, for example, require that particular attention be paid to tool sharpening. The reamer is dry ground on a standard tool and cutter grinder with a grit wheel that removes .001 inch of material at the rate of .0002 inches per pass. Bore and grooves must be absolutely smooth, to a tolerance of .001 inch. The dimensions of both are inspected at every inch along the length of the barrel by means of star gauges, gauges with expanding fingers that transfer their readings to a vernier caliper at the end of a long rod. A spring attached to the star gauge ensures uniform pressure when expanding the measuring points, thus eliminating variations due to the inspector's touch. Fixtures devised to facilitate production included the use of multiple set-ups, quick clamping arrangements, and special indexing devices.

Beretta, when it commenced manufacture of the Garand, had the advantage of hindsight – a decade of experience at Springfield. In electing to build its own fixturing, gauging, and tool systems, the company was effectively blazing a new trail in high-precision manufacture. Although the machines Beretta built could hold the required tolerances, they had to be constantly monitored to prevent excursions from the process limits. This is where statistical process control came in.

The principles of statistical process control acknowledge that machines are intrinsically imprecise, that is, that an identical procedure will produce different results at different times. The degree to which the results vary will depend on the capability of the machine to maintain precision. Sources of variation include a multitude of causes inside

OPERATION SHEET—Rifle Barrel.   1¼-in. dia. steel, SAE 4150

Grind spots for Brinell impression.

Make and record Brinell impression (269 to 311).

Cut to working length 24.200 — 0.010 in. and center ends.

Stamp stock mark on end of stock opposite Brinell mark; end with stock mark to be muzzle end.

Spin on rolls and straighten stock to max. runout 1/32 in. 3½ in. from end of barrel, opposite stock marked end, grind rest spot.

8 in. from end of barrel (opposite end to center rest spot) grind two roll rest spots.

Turn muzzle end 0.895 — 0.020 in. tapered to 1.00 — 0.020 in. at end of cut shoulder of which is 6¾ — ¼ in. from breech end; chamfer muzzle.

Turn end opposite to stock mark to 1.355 — 0.01 in.; from ⅛ in. R. to 0.875 — 0.010 dia.; turn 0.875 — 0.010 dia. straight to 10.237 in. — 0.020 in. from breech end; chamfer breech end.

Spin on centers and straighten to max. runout of 1/64 in.

Inspect 100% (Magnaflux) (after rough turn operation for imperfections in stock construction).

Ream bore to 0.295 + 0.001 in. (Start reamer at breech).

C'sink both ends simultaneously 0.250 — 0.0075 over 0.375 in. ball in muzzle end, c'sink to end of barrel and 0.250 — 0.0075 over 0.375 in. ball in breech end, c'sink to end of barrel.

Spot grind, rest spot for back roll rest 5 3/16 in. from muzzle end.

Turn taper from 0.6751-0.005 in. muzzle end to 0.850-0.01 in. located 6¼ — 1/16 in. from breech end (To remove surplus stock only).

Wash.

Line straighten bore.

Spot grind rest spot.

Turn major thread dia. rough (breech end) 0.985 — 0.005 in. square gage tenon shoulder and 0.975 — 0.003 in. dia. at 1.130 — 0.005 in. dia. shoulder; turn muzzle end 0.625 — 0.005 in dia.

Line straighten.

Spot grind muzzle end, rear of gas cyl. bearing 0.599 — 0.005 in. dia.

Turn taper sections from 0.637 — 0.005 in. dia. located 17.408 in. from breech end thread shoulder to 0.700 — 0.015 in. dia. at lower band bearing. and 0.770 — 0.005 in. at lower band bearing location up to and including ± ¼ in. radius.

Line straighten.

Turn muzzle end 0.540 — 0.005 in. and 0.5813 — 0.0054 in. major thread dia.; form groove.

Burr bore on muzzle and breech end of barrel to permit free entrance for centers.

Chamfer 30° angle on both sides of 8/16-32 P. thd. and form 45° angle on end of gas cyl. bearing.

Finish face gas cylinder lock seat and shoulder of thread.

Grind breech end to 1.115 — 0.001 in. dia. (rough).

Mill thread on breech end and tenon dia., topping nob to produce no larger than 0.972 in. major dia. of thread, 0.917 — 0.002 in. tenon dia.. and 0.913 — 0.005 in. dia.

Form mill top (rough and finish) two at a time.

Wash.

Stamp stock mark and piecemark on stock.

Line straighten.

Grind lower band bearing 0.726 — 0.002 in. dia.

Mill rear hand guard grooves (R & L).

Ring straighten.

Spot grind 0.628 — 0.005 in. dia.

Grind muzzle end (finish) 0.514 — 0.002 in. dia.

Finish grind breech end 1.100 — 0.001 in. dia.

Grind 1 + ¼ in. radius.

Grind gas cylinder bearing 0.660 — 0.001 in. dia.

Mill three cuts for gas cylinder splines, symmetrical and concentric.

Ream chamber (rough) and c'sink breech end 45° to remove surplus stock on breech end of barrel.

Wash.

Rough and finish ramp including 0.01 R + 0.01 in.

Ream bore 0.298 + 0.001 in. dia. (Cut ream oper.) Start reamer at breech end.

Wash.

Ring straighten.

Ream bore (finish) 0.300 + 0.001 in.

Wash.

Rifle bore (hook cut).

Ream chamber (finish) (Cut reamers only used).

Wash.

Broach lower band pin retaining slot.

Form 0.035 + 0.005 R at intersection of ramp and chamber.

Form 0.020 R + 0.005 in. at mouth of ramp.

Mill bullet nose clearance cuts.

Cut thread muzzle end.

C'sink, face and chamfer muzzle end of barrel to finished length 23.310 — 0.01 in.

Hone chamber.

Wash.

Mark manufacturer's initials, month of year and year of manufacture.

Fig. 6.3 Operation sheet for M1 barrel, Springfield Arsenal [3, p 117]

and outside the machine, including causes associated with raw materials, operators, environment, and so forth.

In the language of the SPC era, sources of variation were classified into two categories, random and systematic. Systematic sources are those large enough to stand out, and are associated with specific causes. For example, tool wear leads to slower metal removal each time the tool is used, potentially causing a trend in the dimension from one part to the next. Random sources of variation are the "white noise" that results from a multitude of tiny causes.[1]

If we can observe a process and identify when a systematic error occurs, we can assign a cause and make the necessary correction. This observation can be accomplished by using a control chart (Figure 6.4), which plots observations made of a particular dimension. The control chart is based on the premise that random (i.e., small) sources cause variation that follows a Normal distribution. As such, as long as only random sources are present, 99.5% of the measurements will fall within plus or minus three standard deviation limits of the process mean. When the dimension falls outside the defined limits, there is a high probability that the excursion is due to a systematic error. This is referred to as an "out of (statistical) control" situation, and the operator can then look for a specific cause. Having identified the cause, he can then take corrective action and bring the process back under control. Causes can be very direct, such as a worn tool, or more subtle such as a problem with metallurgy in incoming material.

The control chart is an *attention-focusing mechanism.* It selectively presents, to an operator who has a number of different things to attend to, only those situations that are special and require immediate attention. Introduced by an engineer on the frame line at Beretta, SPC reduced rejects dramatically, from 15% to 3%. How did it accomplish this?

At Beretta, machine reliability was monitored and measured by the mean time between systematic errors. As systematic errors have assignable causes, and as the control chart focuses attention on each systematic error, one is led naturally to manage contingencies in the process. To reduce the mean time between systematic errors one needed to find ways by which the sources of the errors could be eliminated. The application of SPC provided one way by which errors could, over time, be observed, better understood, and eventually solved. Manufac-

---

[1] A clear distinction between systematic and random causes is artificial, although a useful simplification. Today we would say that different problems have different magnitudes and frequencies; the "random" sources are those so small that their sum is Gaussian (Normally distributed).

Fig. 6.4 Blank control chart used by Beretta (bottom half removed)

turing's evolution from an art to a science now included a systematic way of learning by doing.[2]

The success of the method on the frame line led to its use in other areas; over a five-year period, process control charts were introduced at more than eighty stations. All critical operations employed a process control chart and an operator responsible for monitoring the process.

Let us consider the workstation for a moment. With statistical process control, day-to-day management attention was redirected from the quality of a *product* to the performance of a *process*. Concern was no longer with mean performance, but with the "outliers," the out-of-control situations, and not with worker effort, but with machine variance. Inasmuch as semiautomatic machines controlled the pace of operations anyway, the shift in focus from worker effort to machine variance would seem rational. Yet Beretta was unique among small arms manufacturers in making this shift in operations. We will shortly see why.

Semiautomatic machines automate tool movement, thus reducing labor in manufacture. The operator loads the work piece and starts the machine, which then runs unattended until the piece is finished. This provided operator slack time, which could be used either to

---

[2] Although control charts *focus* attention on systematic problems, they are only one of many tools for subsequently *diagnosing* and *solving* the problems. A whole philosophy of problem solving is required. In SPC, this is partly the responsibility of operators, which is a dramatic change from Taylorism. For an extended discussion of the difference between Taylorist and SPC approaches to problem solving, see [23].

monitor the machine or to operate more than one machine. The inclination to increase labor productivity would suggest that an operator be assigned two machines; while one machine is busy cutting metal, the operator could be setting up the second. This was the strategy adopted at Springfield for Garand production.

Such a strategy is reasonable as long as machines are extremely reliable (i.e., the mean time between problems is large). When this is not the case, when we have machines subject to periodic excursions from the process parameters, we not only experience greater rates of rejects and downtime, but contingencies on one machine affect the production output of others. During the early years of Garand production at Beretta, operator slack time was occupied by statistical process control tasks for all critical operations.

If the volume of output is controlled by the speed of the machine and the only controllable element at a workstation is the minimization of problems with the machine, operator attention quite naturally should be on the stability of the process. That processes change over time is implicit in such an ethos of work. *The essence of a process is its dynamic characteristics.* This is in stark contrast to the earlier Taylor view, which cast all problems in essentially static terms. In the Taylor world, there was "one best way" to do something and, having specified it, work was defined by performing it in that way for evermore. In the dynamic view, work is defined in terms of identifying problems and diagnosing and solving them. Supervision of work in a dynamic environment consists not in monitoring effort, but in facilitating change. [23]

6.2. The Quality Control Function

What began as a means for controlling rejects on Beretta's frame line became, over a period of five years, a new system of manufacture. Quality engineering replaced industrial engineering as the dominant ethos. For each phase of the production process the Quality Control Service established:

- points in the production process requiring intervention;

- instruments to be used;

- norms to be followed in identifying and correcting deviations;

- procedures for determining the costs of quality control, scrap, and rework; and

- responsibility and authority of the individual to whom the controls were entrusted.

Quality Control was responsible for quantitatively measuring the natural variability of every machine and the degree of fidelity of every tool, verifying tool conformity to design, and identifying possible causes of systematic error. Through statistical analysis of the collected measurements (Normal distribution and confidence limits) the natural variation of a machine could be calculated and registered and a control method designed for accepted processes.

The quality control system that was established included: frequency of measurement of particular dimensions during manufacture; estimation procedures of process performance, together with methods of diagnosis, if warranted; and procedures for correction of the process by operators, as well as conditions under which machines should be stopped and examined by Quality Control. A quality control supervisor was responsible for eight to ten machines and for day-to-day control of the process in the plant. There were three other staff groups in Quality Control:

- **Testing**, which was responsible for certifying product quality and ensuring conformance to product specifications;

- **Metrology**, which was responsible for the control of tools, the calibration of measuring instruments, and preparation of control instruments; and

- **External Supply Control**, which was responsible for the control of raw materials and partially completed products supplied by outside vendors, as well as for selection of outside vendors.

As might be expected, scrap and rejects were dramatically reduced – from 25% to 8%. But labor productivity improvements were not that significant, 25% over five years, and capital intensity was unchanged. In light of this, one might question our characterization of this change as a revolution.

6.3. From a Static World to a Dynamic One

The intellectual history of process control had seen a shift from a static world view to a dynamic one, in which continual change and improvement have become the *raison d'être* of management. If constant improvement was the focus of management attention, why didn't such improvement translate into productivity improvements? The answer lies in the performance specifications of the products themselves and the introduction of new products with ever-greater tolerance requirements. Consider the Garand. Beretta was able to improve upon this rifle with a new caliber barrel, a new type of magazine and feeding system, and a new sear, called the BM-59, that allowed fully automatic fire upon selection. This weapon was adopted by the Italian Army in three different versions.

Although labor productivity was little changed at Beretta, several characteristics of manufacture changed in fundamental ways. The composition of the work force changed, with quality control becoming an integral part of manufacture and commanding a larger staff. The staff-line ratio went from 60:240 to 100:200. The organization of work changed, with production of each major component organized as a synchronous transfer line. The barrel line, for example, became a synchronous shop with 24 people, nine of them quality control personnel.

A synchronous line is one in which all of the operations required to manufacture a component are rationally laid out, with sequential operations located next to one another on the shop floor. Buffer inventories between operations are kept quite small (equal to one hour or less of work). Throughput time, that is, the time between the start of the first operation on a workpiece and the completion of the last, is greatly reduced by synchronous lines. Because throughput time is short, problems can be caught quickly and corrected before many defective pieces are created. One obvious benefit is that the quality of components is improved and rework is reduced. Another, more subtle benefit is that diagnosis and problem solving are now carried out by examining the workstation not in isolation, but as part of the entire system.

Consider a problem at operation 5 in a process, the assignable cause of which is at operation 3. In a conventional shop with a lot of buffer inventory the prior stations might be working on different batches of products and might have had a new set-up, and the batch of products having the problems might have been made on an earlier shift than the one that detected them. Looking for assignable causes in such a situation would likely be fruitless, as the circumstances under which the problem arose would have changed. In a synchronous line with low buffer inventories, one can examine the entire system for the causes of problems. While this increases the scope of the problem-solving domain, it also involves more than one operator (i.e., a team) in the problem solving effort. A shift in the organization of work has occurred. Now when the assignable causes and solutions for a problem are not quickly visible, a team can be brought to bear.

Together, synchronous lines and statistical process control drive an organization towards an ethos of process improvement that includes a view of an integrated process. Later advances in similar concepts, such as just-in-time (JIT) production in Japan, have amply demonstrated the success and radical transformation of work wrought by such systems. Exception-lot theory is a generalization of JIT that integrates inventory management with SPC for problem detection.[3]

Let us now contrast the principal changes in the nature and organization of work brought about by the introduction of statistical process control with those effected by earlier innovations. First, as we noted earlier in the discussion of Taylorism, the management of work was the management of *effort*. With statistical process control it shifted to management of *attention* for problem detection and solving. It was the "outliers" in a process that were now of interest to management; only by attending to these could one hope to improve productivity or

---

[3] [20] In standard JIT, there is only one buffer between stations. The size of this buffer is normally dictated by production smoothing and lot sizing concerns; the effect on attention signaling is secondary. Exception lots are a second type of buffer whose size is calculated to perform the same function as a control chart, namely signaling only when a systematic error occurs.

quality. Problem solving, a cognitive ability, stressed information processing and diagnostic skills. Learning about outliers could only take place when problems were recognized, described, and solved and this, of necessity, could only take place at the workstation with the help of the operator. Discretion and control of work, which were earlier removed from the operator, were restored.

Second, under Taylorism, besides the obvious task of making sure procedures were executed as planned, the principal management task was coordination. With functional specialization of labor, one had to be a concertmaster to ensure that the firm, as a whole, functioned efficiently. The planning center, which served this coordination function, was the nerve center of productive activity. With SPC, the quality control department took over the principal functions of manufacturing management. Concern for schedules and production output was subordinated to concern for quality and process improvement.

Third, synchronous lines forced an integrated view of the entire system of manufacture. Whereas the intellectual underpinnings of Taylorism were *reductionism* and *specialization,* that of SPC in a synchronous line was *integration.*

Fourth, information management of process parameters was institutionalized. With the introduction of SPC there was, for the first time, explicit recognition and separation of information about operating process parameters from the physical processing of material. The separation of information about a procedure from the procedure itself was the intellectual watershed crossed by statistical process control. Now it was possible to observe and study the efficacy of procedures.

It would not be very long before one could completely separate information processing from material processing. This would come in the form of the next technological breakthrough, numerically controlled machines.

7

The Numerical Control Era

Beretta acquired its first numerically controlled (NC) machines in 1976. These machines functioned automatically, performing operations and changing tools according to numerically coded instructions. Although this technology had begun to spread through Italy at the beginning of the decade, its presence was isolated. Its primary users were companies that manufactured small quantities of products of high value. With the introduction of microprocessors, controllers went down in cost and up in reliability, making NC technology viable for large scale use. Beretta introduced these systems into the high volume production (200 to 400 pieces per day) of small- to medium-size products.

Beretta regarded the automation of tool changing as the single most significant benefit of NC technology. Automated tool changing meant that what had formerly required a transfer line could now be accomplished with a single machine. NC machines (see Figure 7.1), which combined the versatility of general-purpose machines and the productivity of special-purpose machines, also overcame limitations imposed on particular components by the specialization of transfer lines. But they were expensive. At the time, the best Beretta was able to obtain was a four-year payback, and some had an eight-year payback.

"As you can see," said company president Ugo Beretta, "it was not what you would call a brilliant investment. But we had to do it sometime. We could have waited, but we could not turn back the clock. It was a very new technology, with electronics and computers, and we had to understand it. Instead of waiting, we decided we would go ahead and buy a machine tool company and learn the new technology. So we bought MIVAL, a small machine tool company with expertise in this field."[1] The effects of numerical control at Beretta are summarized in Table 7.1.

Numerical control had evolved out of a program funded by the United States Air Force in the late 1940s for making complex shapes. Although the first commercial products were offered a decade later, there was no significant penetration of NC systems until controllers became more economical and reliable in the 1970s. Although self-directed machines – automation – went back many decades, a critical distinction between NC and earlier automation was that the sequence of tasks could be easily altered or replaced by loading a new program.

The work cycle of NC machines – the set of motions that determines the selection of tools, their proper positioning in three dimensions relative to a workpiece, feeding of workpieces, flow of coolant, and so forth – was recorded as a series of codes initially on punch tapes, then on magnetic tapes. This information, called a part program, was passed to the "programmable controller," a crude special-purpose computer that processed the information and issued signals to the various motors on the machine to position the machine axes accurately and precisely, and cut the workpiece. The motions a machine tool must go through to produce a part must be described in detail, mathematically. This reduces the entire process of producing a part, including the skill of the machinist, to a formal, abstract expression, which, when coded and translated by a microprocessor, activates a machine's controls. Every machine movement, however slight, has to be formally, explicitly, and precisely articulated. With such programmable automation, a switch in products no longer entails physical set-up changes to retool or readjust the configuration of the machines, only a switch in programs. Thus, NC technology combines the versat-

---

[1] Personal communication.

	Summary of Epoch	Numerical Control (NC)
Size Trends	Introduced at Beretta (world)	1976 (1960)
	# of People (Min. Scale)	100
	Number of Machines	50
	Productivity Increase (over previous epoch)	3:1
	Number of Products	100
Nature of work	Standards for Work	Functional standards
	Work Ethos	"Control"
	Worker Skills Required	Experimental
	Control of Work	No supervision of work
	Organizational Change	Cellular control
	Staff/Line Ratio	50:50
	Line Workers per Machine	1
Technology Keys	Process Focus	Adaptability
	Focus of Control	Product/ process integration
	Instrument of Control	Electronic gauges
	Rework (as fraction of total work)	.02

Table 7.1 Effects of Numerical Control at Beretta

ility of general-purpose machines with the precision and control of
special purpose, or self-acting, machines. [31]

"In the past," observed *The American Machinist* in 1973,

> humans were both translators and transmitters of information:
> the operator was the ultimate interface between design intent, as
> incorporated in a drawing or instruction, and machine function.
> The human used mental and physical abilities to control machines.
> Today, computers are increasingly becoming the translators and
> transmitters of information, and numerical control is perhaps most
> representative of the kind of control that plugs into the greater
> stream with a minimum of human intervention. Historically,
> numerical control certainly has been the most significant develop-
> ment of the electronic revolution as it affects manufacturing.
> Quoted in [29, p 221]

CENTRO DI LAVORO ORIZZONTALE   HORIZONTAL MACHINING CENTER
MI-VAL 026 FMS   MI-VAL 026 FMS

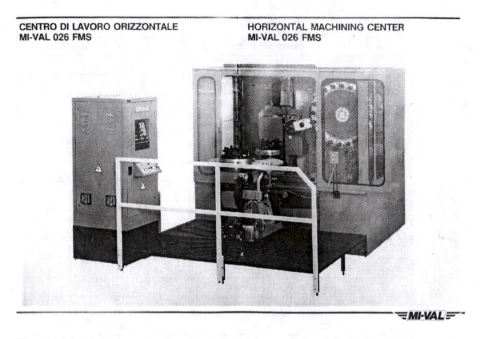

Fig. 7.1 Machining center; the circular structure on the right holds different tools,
which are automatically selected during operation (Source: advertising brochure,
Pietro Beretta)

NC technology, after two decades of disappointment, came into its own with the advent of microprocessors. Microprocessor technology made controllers at once extremely powerful and relatively inexpensive and its greater computer power made possible sophisticated, yet flexible and "user friendly," operator interfaces.[2] It also made possible advanced control techniques including allowing NC machines to record utilization and cutting tool life, reduce set-up efforts and time, compensate for errors, inspect surfaces and make automatic adjustments, allow operators to modify their programming on the shop floor, record events of the last minute or two prior to a failure, and perform self-diagnosis. Coupled with greater sophistication in machine tool design, numerical control using microprocessors made possible the development of stand-

---

[2] This is often referred to as computerized numerical control (CNC). In this section we do not distinguish the two, but view them as parts of a continuum of increasing sophistication in numerical control.

alone machining and turning centers capable of shift-long, untended operation.

The early problems of NC technology were partially due to limited formal knowledge of the machining process. A lot of the knowledge possessed by skilled machinists, such as when and how to make "on the fly" adjustments, was tacit or otherwise not accessible to programmers.[3] This limited understanding of contingencies and variations in factors such as machinability, tool wear, and part material properties significantly constrained early implementations of NC technology. But with effort, over time more of the tacit knowledge implicit in operator skills became precise, explicit knowledge that was used to develop procedures capable of avoiding or dealing with a variety of contingencies.

7.1. NC Technology at Beretta – From Synchronous to Cellular

What happened to the organization of work in the Beretta plant after the installation of NC machines is interesting. In the transfer-line the average cycle time for a product was two minutes. Half of this was attributable to the machine, the other half to the operator. With the automation of tool changing, a variety of operations could be done by a single machine, but the overall cycle time increased. The cycle time required by an equivalent NC machine to perform the operations that previously required three machines would be 3.6 minutes, only .6 of which would be operator time. Thus, one NC machine replaced three machines, but took almost twice as long to produce a single part.

One can see that an operator of this NC machine would be idle 85% of the time (3 minutes out of 3.6 minutes). By allocating two machines to each operator, he would be busy 1.2 minutes while the

---

[3] For example, a potential problem in machining is "chatter," a forced vibration of the tool against the workpiece. With conventional machining, it is "easily detected by an operator because of the loud, high-pitched noise it produces and the distinctive 'chatter marks' it leaves on the workpiece surface." [24] Once detected, an expert machinist can halt it and even correct the surface finish on the fly, if necessary. But this was far beyond the ability of early NC tools. In its absence, the machining procedure must be programmed conservatively to avoid potential chatter. See [7].

cycle time would still be 3.6 minutes. Thus his idle time could be reduced to 66%. The greater the machine component of the cycle time, the larger the cluster of machines it makes sense to put around the operator. This leads to a cellular rather than synchronous plant layout.

In a synchronous line with two-minute cycle times, an operator performed a fixed, unchangeable routine. The nature of the work was determined by "hard automation," the jigs, fixtures, and cams that governed the performance of the operation. With hard automation, considerable effort was expended to get the jigs and fixtures right the first time. "Quality" was front-end loaded in the hardware design and quality control was a process of monitoring and tending the machines and tools.

The scope of activity at any given workstation was very small and the machine established the pace of work. The principal intellectual activity on the line consisted in monitoring machine performance and diagnosing problems when they occurred. Because a problem at any one station on a synchronous line could stop all subsequent operations, thus exacting a high cost in productivity, a large and centralized set of resources was allocated to problem solving. At Beretta this allocation was seen in the growth of the Quality Control department.

A cellular plant layout significantly increases the scope of activities for which an operator is responsible. The twelve operator stations in the barrel line layout shown in Figure 7.2 are responsible for one hundred and sixty-eight operations, an average of fourteen operations each. This compares with an average of three operations per person on a typical indexing machine in a transfer line used to manufacture, for example, the Garand rifle. Thus, we have a five-fold increase in the scope of activities.

We find, too, that the nature of the work changes. An NC operator works not with physical objects, but with information. The object of attention and medium of work is a computer program. Whereas the operator on a synchronous line was interested in observing the behavior of a *process*, the operator in a manufacturing cell composed of NC machines is interested in observing the behavior of a *procedure*.

Fig. 7.2 Beretta barrel line layout under NC (Source: company documents)

7.2. Softening "Hard" Automation

NC machines are characterized as "soft," or programmable, automation. Operators write programs that precisely specify, down to the most minute detail, a sequence of operations that involve a choice of tools, the length and direction of tool movement, and the speeds and feeds of machine controllers. These sequences are contingent on measured conditions such as tool wear and compensation.

Soft automation possesses five distinguishing characteristics.

- **Specificity of procedures**. The degree of detail with which a procedure must be specified is at least one order of magnitude greater than with hard automation. The number of lines of program code required to machine a typical barrel was 6,300. For every possible contingent condition, such as variations in the dimensions of a raw casting, we need a response in the form of a clearly defined procedure. Because the computer is static and functionally blind, capable only of moving a tool in three dimensional space and changing its course at prescribed points, the procedures must be written as if to guide a blind person restricted to a small set of activities in a finite space. The specificity of the procedure, together with removal of the person from the immediate environment of work, renders the activity more abstract and scientific.

- **Adaptability to change**. Procedures are changed and new procedures implemented by making changes to existing, or writing new, computer programs. Hence, quality is no longer front-end loaded, but subject to constant change and improvement that can be observed, monitored, and modified at the workstation. Because change does not entail the design and construction of new hard automation, we see more frequent and incremental changes in procedures that do not require centrally allocated resources being introduced at the workstation. Thus, work at a station now involves not only monitoring performance, but improving it as well.

- **Versatility of operations**. Operations at a workstation are only restricted by the configuration of the part being machined – whether it is prismatic or rotational. NC machines can perform operations on either one of these two classes of products, but not on both. Within each class, the machines can perform almost any operation, restricted only by the availability of tools in the tool magazine and the tolerances the system is capable of maintaining. This avoids the large fixed cost associated with special-purpose machines by enabling a variety of new products to be machined at a single workstation. Thus, the scope of activities at an NC workstation is expanded to include the introduction of new parts and processes. Precision, adaptability to change, and versatility of the machines have rendered the nature of work at a workstation more scientific and abstract, more comprehensive, and subject to continual change.

- **Reproducibility**. Once a program is written, the machine controller is capable of executing the program consistently forever. This suggests that the better able a program is to deal with contingencies in operation, the less need the machine will have for a skilled operator. An operator writing a procedure is, in effect, "cloning" him or herself. If the cloning is perfect, the operator is left with no job at all, or at best a very uninteresting job. This phenomenon creates a managerial imperative to constantly introduce new products and processes in order to keep skilled people in the organization occupied or suffer possible disintegration into an organization without the skills to innovate.

- **Transportability**. A reproducible program's use is not restricted to the machine on which it was developed. It can be used on any identically configured machine and it can be copied at almost no cost. Thus, once a program is written parts can be subcontracted to any small job shop with equivalent equipment without a great deal of concern for quality. Quality is assured by the raw materials and the programs that govern the parts' fabrication.

Reproducibility and *transportability* permit the scale of a manufacturing enterprise to be small and assure the growth of the enterprise without the addition of skilled people. The five characteristics of information intensive processing – precision, adaptability, versatility, reproducibility, and transportability – suggest a complete restructuring of the organization of work and the nature of the firm.

Do we find such changes in practice? The substantial impact of its NC machines on quality, and the concomitant enhancement of the quality control organization, were benefits that Beretta had not fully anticipated. Management of quality with transfer line technology, with engineering and manufacturing separated, had consisted in monitoring and control. With NC technology, management of technology grew to encompass methods engineering and moved manufacturing a giant step closer to a science.

With its base of experience in automation, Beretta decided, when computer numerically controlled (CNC) machines became available, to completely convert its machine tool base to this newer technology within six years. Abetted by cost reductions that accompanied the spread of this new technology, Beretta had, by 1984, installed more than two hundred CNC machines on which more than 90% of its metal work was being done. Total value added cost in manufacturing, due to better quality products and substantially less overhead, was reduced from sixty-six cents on the dollar to sixteen cents on the dollar.[4]

---

[4] This sentence is ambiguous, but the change is clearly dramatic and important. "Value added cost" is not a standard term. It probably means that manufacturing overhead as a percent of final output fell from 66% to 16%. Accounting systems differ, but apparently rework and defects were charged to overhead.

7.3. The Impact of Numerical Control on the Quality Organization

To understand the impact of NC machines on the organization of work one needs only look at the changes in the quality organization. The object of attention, as noted above, is no longer how a process behaves, but rather how a procedure behaves. Increased microprocessor control of activities and on-line analysis of tool wear and tool compensation provided automatic feedback, enabling cybernetic control of machining. With control of the process automated to such an extent, the nature of quality management was bound to shift.

Figure 7.3 shows an electronic gauge used for some turning operations at Beretta. An operator introduces machined parts into the gauge, which measures four different dimensions. The measurements are automatically fed into the FANUC controller, which integrates them with data on parts previously produced. The NC machine has some built-in variability and the sensor has measurement error. We would expect that parts would not be identical, but would range randomly within certain bounds of precision. We need a procedure that, taking account of this random variability and the historical data, can detect a systematic change (jaws misaligned) or a trend (tool wear). Having detected a change, we need to make an adjustment to the cutting program. The machine then automatically adjusts the appropriate tool movement.

In order to have the machine make the adjustment automatically, it is necessary to understand all the steps in the cutting operation. The step at which the adjustment is made must be appropriate as a point of intervention to accommodate a change and the kind of change must be appropriate to the procedure. Statistical sampling, techniques of inference, and methods engineering are all integrated within one procedure.

We can see, in the example above, that *methods engineering and quality control have become one and the same*. The industrial engineering of the Taylor era and the quality engineering of the SPC era have now been subsumed in a new discipline, *systems engineering*. In 1979, Beretta formalized systems engineering as the Quality Control Programming Section and charged the head of the section with responsibility

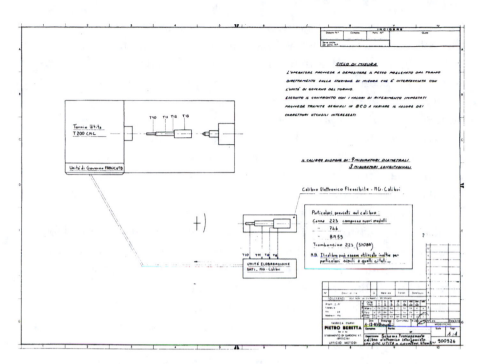

Fig. 7.3 Electronic gauging plan

for all of its manufacturing procedures. Data analysis, experimentation with new procedures, and evaluation of new technology, as well as documentation of all that went on in manufacturing, fell under the purview of this section, which grew quickly to 22 people, becoming the largest group in the quality control organization.

The group was charged with analyzing the individual phases of a production process (machines, operators, work methods, and working conditions) to ensure their conceptual fit in order to guarantee quality and to determine whether the preparation of each phase was consistent with overall project development.

The group was also charged with analyzing all available means of production for the purpose of quantitatively measuring their natural variability (machines) and level of fidelity (tools), and for eliminating, for every operation, every possible cause of systematic error by: verifying tool conformity to design (metrology); testing with a tool that precisely met design specifications; and producing pieces in a quantity such that tool wear would not degrade quality. Through statistical analysis of the collected measurements (distribution and confidence limits), the natural variability of a machine could be calculated and registered, a specific procedure could be accepted or rejected on the basis of its ability to maintain the required tolerances, and a control method could be designed for accepted processes.

With on-line, 100% inspection, the inspector should be at the workstation where diagnosis of problems takes place based on information derived from every part in production. This suggests that the quality control and methods engineering functions are now being done at the workstation by the operator. As we can see in the barrel line illustration, each of the 12 workstations has a quality control bench. The distinction between line and staff is sufficiently blurred by this shift as to render arguable whether the line-staff concept is still meaningful. On this line each operator is both quality inspector and methods engineer. The operators, who formally report to the Quality Control Programming Section, are responsible for, and have authority to make changes to, procedures.

The shift to managing information and procedures was not an easy one for Beretta. "It was," averred Ugo Beretta, "the biggest change in the culture of the plant that I have ever seen." Each operator is the manager of a cell, and there is no supervision of work. Operators no longer monitor the performance of machines, but rather control the performance of a group of machines run by computers. To do so they need to understand the relationship between computer programs and physical output. They also need to understand the interaction of all aspects of a system of machining. The principal medium of communication is no longer a blueprint, but printed output.

The use of the electronic gauge (Figure 7.3) is a particularly telling observation in manufacturing practice. There are no tolerance or measurement specifications, only four parameters labeled T10, T11, T12, and T13. These specifications are replaced by control programs, each of which represents a relationship between some set of historical measures and the required response. To specify these control programs the operator has had to become a systems engineer, comfortable with database manipulation and control algorithms. The transformation of work has been quite radical indeed.

How did this transformation affect operations at Beretta? For a start, the company's output grew threefold in eight years without a net increase in the work force and was still capable of handling excess capacity. Having drastically reduced manufacturing costs, Beretta began to manufacture rifles for its two major competitors, Browning in England and FLN in Belgium. In 1985, the Beretta Parabellum 9mm won the hotly contested contract to replace the venerable Colt .45 automatic pistol for the United States military. The contract stipulated that a new factory be constructed in the United States. The reproducibility and transportability of its NC programs assured Beretta of being able not only to make money at a bid price less than half that of the second place bid, but also to start up an entirely new factory in Accokeek, Maryland within eighteen months.

NC technology simultaneously enhanced flexibility, quality, and productivity. Beretta realized a tenfold increase in the number of products that could be produced on the line, with a concomitant reduction in rework and scrap from 8% to 2% and a threefold increase in labor productivity. Implicit in the simultaneous increase in number of products and quality of workpieces is an integration of product and process knowledge.

The workplace ethos had changed again. It was now more than just monitoring machines; it was controlling them as well. Electronic gauges replaced control charts. The skill required was more than diagnosis. One had to experiment with procedures and learn. Adaptation replaced stability as the process focus. System engineering replaced quality engineering as the dominant engineering ethos. Information technology had come of age.

8

Computer Integrated Manufacturing – The Dawning of a New Age

Just as Beretta completed the renovation of manufacturing machinery in its plants, yet another new technology began to emerge. Robots for loading and unloading parts in machines, untended mobile carriers for transporting pallets from one part of a plant to another, and flexible manufacturing cells capable of a tenfold increase over traditional machinery in the variety of parts that could be made were all making their debuts, and with them came the potential to automate the manufacturing process from one end to the other, from loading machines, through changing, setting, and operating tools, to unloading processed parts.

In 1987 Beretta engineers introduced, as pilot projects, two new technologies: a flexible manufacturing system (FMS); and computer-aided design/computer-aided manufacturing (CAD/CAM, the CIM integration of computer-aided design and CNC machines). CAD/CAM eventually became Computer-Integrated Manufacturing (CIM). The effects at Beretta are shown in Table 8.1 .[1]

---

[1] Editorial note: From the vantage point of 2004, computer-integrated manufacturing has become ubiquitous and overwhelmingly important in manufacturing, whereas FMSs are in a limited niche and likely to remain there, partly for reasons discussed later. Jaikumar was initially more interested

Summary of Epoch		Computer-Integrated Manufacturing
Size Trends	Introduced at Beretta (world)	1987
	# of People (Min. Scale)	30
	Number of Machines	30
	Productivity Increase (over previous epoch)	3:1
	Number of Products	Infinite
Nature of work	Standards for Work	Technology standards
	Work Ethos	"Develop"
	Worker Skills Required	Learn/ generalize/ abstract
	Control of Work	No supervision of work
	Organizational Change	Product-Process Program
	Staff/Line Ratio	20:10
	Line Workers per Machine	0.3
Technology Keys	Process Focus	Versatility
	Focus of Control	Process intelligence
	Instrument of Control	Professional workstations
	Rework (as fraction of total work)	.005

Table 8.1 Effects of Computer-Integrated Manufacturing and Flexible Manufacturing Systems at Beretta

8.1. Beretta's FMS

A flexible manufacturing system is a computer-controlled configuration of semi-independent workstations, connected by automated material handling systems, designed to efficiently manufacture more than one kind of part at low to medium volumes. Beretta's first project was the installation of a flexible manufacturing system for manufacturing a major gun part, the "receiver." The system designed for production of the Beretta receiver (Figure 8.1) consists of three CNC machining

---

in FMS, however, because it corresponds to removing operators *entirely* from normal operations, up to and including a lights-out factory that still has versatility. This requires another order of magnitude improvement in the science of manufacturing, to achieve good up-time and quality with no possibility of human intervention. I have left the original discussion of both topics essentially unchanged, but added a few footnotes on subsequent developments.

centers connected by a material handling system that incorporates a conveyor arranged in a loop. The loop constitutes a buffer area; pallets on which the workpieces are mounted keep moving until the machine required for the next operation becomes available. The system is capable of fabricating forty-five discrete parts. With the exception of inspection, all system operations are under computer control.

In most FMS installations incoming raw workpieces are hand fixtured onto pallets at a workstation. Once information on a fixtured workpiece (typically an identifying number) has been entered to inform the FMS that it is ready, the FMS supervisor (supervisory computer) takes charge, performing all the necessary operations to completion in any of a number of machines, moving workpieces between machines, responding to contingencies, and assigning priorities to the jobs in the system.

The supervisor first sends a transporter to the load/unload station to retrieve the pallet. The loaded pallet then keeps moving in a loop until a machine becomes available to perform the first operation. When a shuttle (a position in the queue) is available, the transporter stops and a transfer mechanism removes the pallet, freeing the transporter to respond to the next move request.

Parts received by the machine must be accurately located relative to the machine tool spindle. The inspection to accomplish this can be done manually, using standard instruments, or by coordinate measuring machines. The appropriate machining offsets are calculated from the measurements and communicated to the supervisor.

Meanwhile, the supervisor has determined whether all of the tools required for the machining operations are present in the tool pocket of the machining center, and requested needed tools from either off-line tool storage or a tool crib/tool chain within the system. When all the required tools are loaded, the supervisor downloads the NC part program to the machine controller from the FMS control computer.

The process of making sure that the part is, in fact, what the computer thinks it should be is termed *qualifying* the part. Qualifying includes making sure that all previous operations have been completed, that the part is dimensionally within tolerance limits, and that it is accurately located. Tools, too, must be qualified. Tool geometry, length,

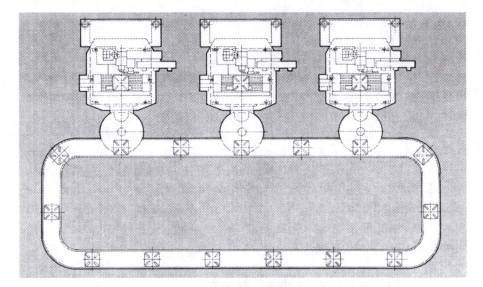

Fig. 8.1 FMS line for receiver production

diameter, and wear are all examined, either manually or under computer control. When both the workpiece and the tool have been qualified, the tool, part, or program offsets necessary to correct for systematic error have to be established.

When the set-up activities are completed, machining begins. The FMS monitors the tool during machining. If it breaks, a contingent procedure is invoked. Some advanced FMSs have in-process inspection and adaptive control whereby a continuous measurement of metal removal is taken to determine whether the operation is within defined process parameters. Compensating corrections for any deviations are made during machining, without stopping. Adaptive control in FMS is still very rudimentary and technically quite difficult with [late 1980s] technology.[2]

The finished, or machined, part is moved to the shuttle to await a transporter. After being loaded onto the transporter, the pallet is moved to the next operation, or else circulates in the system or is

---

[2] Considerable progress on adaptive control has been made since the 1980s in both sensor technology and adaptive control algorithms.

unloaded at some intermediate storage location until the machine required for the next operation becomes available.

The computer controls the cycles just described for all parts and machines in the system, performing scheduling, dispatching, and traffic coordination functions. It also collects statistical and other manufacturing information from each workstation for reporting systems. As all the activities are under precise computer control, effects of part program changes, decision rules for priority assignment, contingent control, and part-portfolio mix can be captured, at least in principle.

The pre-FMS line layout for making receivers is shown in Figure 8.2. The 41 machines in this line compare with the FMS line's 24, configured as eight parallel three-machine cells (the number of cells dictated by the volume of work). Each cell in the FMS receiver line fabricates a complete receiver and is managed by a single worker. The FMS reduces minimum efficient scale by an order of magnitude, from 41 machines to three, and is flexible and versatile enough to accommodate other prismatic parts as well as receivers.

It will eventually be possible to load a machine on the FMS line at the beginning of a shift with thirty-five pallets, each containing a blank receiver, and have the entire lot completely machined by the end of the shift without an operator being present. Although untended operation has not yet been achieved at Beretta, it is not only possible, as a number of Japanese machine tool vendors have shown, but achievable in the next decade. [19] When it comes it will in all likelihood once again radically alter the nature of work.

What we can expect from a world of untended flexible manufacturing is summarized below.

- The worker is likely to be completely separated from the physical elements of work – metal, lubricants and oil, executing procedures, and turning out parts. Work will, instead, become an act of conception, of creating new products and processes.

- All of the tools, fixtures, and programs needed by a system will have to be conceived, built, and developed before it can make the first product. Thus, all of the controllable costs will be sunk before the first product comes off the line, after which

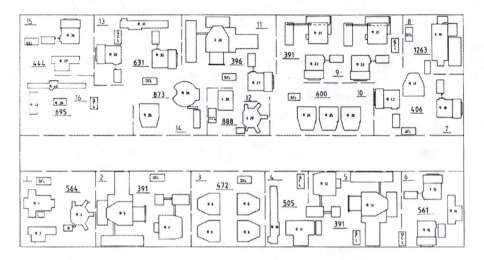

Fig. 8.2 Beretta receiver-line layout before FMS

the unit cost will be the same whether the firm makes one unit or many.

- In order to achieve untended manufacture the craft of machining needs to be developed into a science of manufacturing. Every possible contingency needs to be anticipated and an appropriate response provided in the form of a tightly specified procedure.

8.2. Knowledge and Problem Solving in FMS [3]

With each epochal change, from Statistical Process Control to Numerical Control to Flexible Manufacturing Systems, the necessary knowledge became more extensive, more formal, and at a higher stage. Concurrently, the process of problem solving, which generates much of that knowledge, also had to change. The reason is that with each epoch, line operators got farther away from the physical elements of work, so that their intuitive pattern recognition and expertise were not accessible. In the SPC era and before, master mechanics working with

---

[3] The concepts in this section are discussed more extensively in [31].

general purpose machines usually accrued years of experience, during which they accumulated a wealth of idiosyncratic knowledge about how to perform in a wide variety of circumstances. They talked in terms of a "feel" for the machine, the tools, and the parts they worked on. It was through this feel that they were capable of producing parts to exacting specifications. Watching them work, one had a sense that they recognized errors (e.g., vibration, chatter, structural deformation due to thermal forces) as they were happening and adapted their procedures to compensate for them. This, in engineering terminology, is an advanced form of adaptive control in an ambiguous environment. Such adaptive error recognition and compensation requires either very elaborate expertise with a complex web of relationships, such as the experiential and partly tacit knowledge of the skilled machinist, or alternately a high stage of formal knowledge approaching full scientific understanding of the machinery, sensor, and controller technology, as well as of the product, the process, and all their interactions.

With the advent of numerically controlled machines, the master machinist was often replaced by a less skilled operator. This does not imply that contingencies were somehow removed from the machining process. All the new machines did was follow explicitly well defined procedures in the form of computer programs. Yet, dynamic contingencies remained a part of the environment, and skill was still required to identify and eliminate errors. Neither the computer control systems nor the lesser skilled persons operating them were capable of diagnosing systematic errors in these machines. The "feel" for the machine was absent. New skills, those of manipulating abstract procedures and entities and recognizing and learning from the relationships between procedures and outcomes, were required.

FMS technology and "unmanned" machining compounded the problem of dealing with contingencies. Workers are entirely removed from the machining area, machining being done using multiple machines with multiple tools and inspection done off the machine. As an FMS is merely a number of standard NC machines connected by an automated material handling system, it has all of the problems common to NC machines. But it also lacks the stand-alone NC machine's almost constant attention from a machine operator, who can compensate for small machine and operational errors by realigning parts in a fixture, tweaking cutting tools, visually inspecting parts between workstations, and so forth. When the final result is wrong, determining the source of the problem in an integrated FMS can be very difficult. The error might be the result of any one or a combination of factors, as discussed below. Proper diagnosis entails knowing which tools were used on which workpieces on which machines and, if more than one part program was used, which was in use when the problem occurred.[4]

The level of complexity involved in problem solving on an FMS is an order of magnitude greater than in a manually tended NC machining center. Thus, if an NC machine is once removed from the "feel" of machining, an FMS is twice removed. To understand the difficulty associated with diagnosing problems in an FMS an analogy is useful. Consider the task of writing, for a person of limited vocabulary and using only the English language, the instructions for drawing a picture of a donkey. If this exercise proves easy enough, then consider the following: each of three people is to be given instructions for drawing a different part of a donkey (using different vocabulary and syntax) and their drawings are to be brought to a central location where a fourth person will be instructed to assemble them. If a fifth person were to inform you that the donkey looks like a horse, how would you go about correcting the problem and issuing new instructions? In order to move from an art to a science, we need to understand the streams of knowledge that make up the science. It is not the case that the problem solvers are no longer skilled. In fact, they are highly skilled; but the domain in which the skill operates is different, having become more abstract. Workers in an untended FMS are "staff" responsible for development, rather than "line" responsible for day-to-day operations.

Why, you might ask, if problem diagnosis is so difficult, are FMSs used to fabricate high accuracy parts? It is because of controllability,

---

[4] As the sequence of workpieces, tools, work centers, and programs used to make a part is controlled by the FMS, in principle complete information is available about what was done to each part, but in practice software tools did not exist to make sense of the huge amount of data generated each shift. Development of factory information systems, to make sense and take advantage of the data generated by complex microprocessor controlled tools, has been a major area of progress in the 1990s and beyond.

reproducibility, and reprogrammability. Once we have solved the problems and written the appropriate code, the system will reproduce the procedure forever. Reprogrammability allows us to perform experiments on the line to correct for errors. If the procedures are set up right, codification of the experience gained and of alternatives taken and rejected becomes possible. In a restricted domain such knowledge can be transferred to other processes, products, tools, and so forth.

Where a number of contingencies can arise it is important to be able to recognize, diagnose, and learn from the resulting errors, and then to generalize from the specific problem to anticipate and avoid similar problems before they occur. With such a high plane of technical knowledge required, operators have to be trained in the scientific method in order to better understand how various machine tool errors can cause parts to be out of tolerance, how to measure and correct these errors, and how to make accurate parts.

An operator is usually alerted to possible problems through discovery of an error in the final form of a part, such as size, shape, location, or surface finish. Errors can result from one or a combination of disturbances in three broad classes: mechanical, thermal, and operational. These are elaborated in Figure 8.3. Both mechanical and thermal disturbances can be further classified as attributable to aspects of either a machine or a part.

The pattern of disturbances can also be important. They can be systematic (static), whereby they reoccur with approximately the same magnitude each time the manufacturing task is performed, or they can be random (dynamic/fluctuating), occurring at random times with different magnitudes and without an apparent pattern. Many systematic disturbances can be avoided by good shop and machining practices, maintenance discipline, and an awareness of how fixture design, poor tool setting, and other actions can affect system performance.

When process capability is high, meaning that process variation is small relative to required tolerances, error avoidance is enough to keep a shop running smoothly. Exploration of machine tool errors and diagnostics do not play a significant role in day-to-day operations and these skills will not be required of shop floor personnel. But in a changing environment, in which new parts and part programs are being

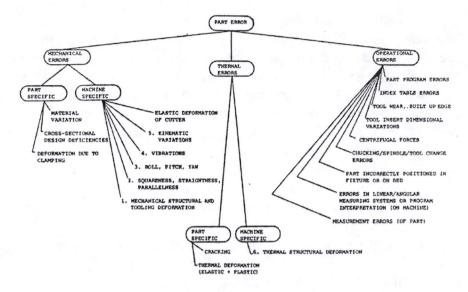

Fig. 8.3 Classification of errors in an FMS (Source unknown)

introduced and tolerance bands are tight, error avoidance alone is not likely to prevent defects. Detailed examination of machine tools is required to locate sources of mechanical and thermal disturbance, determine their magnitude, and identify their mechanisms. It is possible to segregate these into recurrent and transient disturbances through a number of tests. Systematic errors are significantly easier to deal with than random ones. If, for example, a machine tool has a bed that sags slightly at one end when the table is moved over it, and if only one axis is affected, it might be possible to reprogram offsets in another axis to compensate for the sag while cutting.

Random disturbances, whose sources are usually thermal, are the most difficult to handle, requiring extremely careful analysis of alternative procedures. Their solution requires the "feel" of a master machinist, the logical mind of a software programmer, and access to extensive databases of information on similar problems.

With this transformation of the operator into a "knowledge worker," the blue-collar image of factory work is no longer appropriate. The intellectual assets needed by a firm to create new products and processes become the dominant driver in manufacturing competence. Thus, management of these assets becomes crucial to a firm's viability.

8.3. Beretta's CIM System

An understanding of what these assets are can be gleaned from an analysis of the second project Beretta undertook, integration of its CAD/CAD system. This project consisted in having both the design for the locking system of a rifle and the NC programs needed to directly fabricate its parts developed on an engineering workstation. The three engineers assigned to this task were provided with a workstation with a color graphics terminal that displayed icons intended to facilitate component design. The component design had to satisfy both product functionality requirements such as safety, efficient kinematic interaction, ability to withstand stress, and ease of assembly and disassembly, and manufacturability constraints such as required tolerances, clearances between components, simplicity of parts configuration.

The locking system comprised 26 different parts. Some of these were selected from a catalog of components; others had to be specifically designed. Software vendors supplied a number of different computer programs capable of manipulating parts geometries. These were used to create designs, move them around on the screen, and fix them to specific locations as if one were actually physically assembling them in three-dimensional space. Other computer programs simulated the movement of the triggering mechanism and the kinematic linkages associated with it. The forces exerted on the mechanism during firing and the resulting stresses were also simulated by computer programs.

An engineer working with a number of different parts geometries could create and test different alternatives, settle on a tentative design, and then examine the manufacturing impacts of each part. A host of manufacturing related computer programs could then be used to create the NC programs needed to machine the components and even graphically display the tool path of a metal cutting program on the screen. When satisfied with the design, the engineer could transfer the program to a machining center and have the components fabricated automatically.

The CIM pilot project proved that the concept was viable. Three people designed in as many months what normally took nine people a year to develop, a 12:1 improvement.

The components used to test the feasibility of the concept were, by design, quite conservative, but the engineers realized that by building a variety of analytical models they could create a very powerful design tool. By simulating performance in real-time they learned much about how the different aspects of product design and process interacted. They were able to discern problems in these interactions and even to codify design rules for manufacturability.

The immediate organizational response to the CIM project was to regard this design tool, like all the other innovations the company had adopted, as a productivity enhancement tool. This it was, but it was also something more; it was a knowledge enhancement, or *learning*, tool. The system possesses the potential to make an expert more of an expert and, as it accumulates information, models, and design rules, to enhance the intellectual assets of the firm. It can be argued that such systems are themselves part of a firm's intellectual assets.

If the enhancement of intellectual assets is critical for manufacturing competence, and if CIM integration has the ability to achieve it, in what dimensions might we discover useful insights? Obvious dimensions are those of organizational memory and analytical capabilities. CIM systems are capable of maintaining, and providing on-demand access to, vast stores of information and applied knowledge. They can perform a variety of calculations and simulate behavior to reduce uncertainty in product and process performance. Taken together, these capabilities constitute powerful productivity enhancement tools for both design and manufacturing.

A particularly important dimension that such tools add to a firm's intellectual assets might be called system intelligence. Prior to CIM integration, organizations solved product/process problems by taking recourse to the respective experts in each area. These experts are repositories for vast stores of functional knowledge. Over time, organizations have learned how to effectively manage the knowledge of these functional experts and devised mechanisms for resolving the conflicts that arise between them over which sets of alternatives are better in a given situation. This knowledge of how to manage knowledge is sometimes called *organizational intelligence.*

In CIM integration we have begun to include information on, and models of, functional expertise. This information enables us to systematize, examine, and learn from interactions between functions in such a way that issues are more sharply focused and patterns of interaction become recognizable. *System intelligence*, then, is the recognition and understanding of the interactions between functions and a surrogate for organizational intelligence as it relates to product/process performance.

Creating functional models of products and processes, validating them with experience, and manipulating them in the process of design is an emerging form of engineering called *knowledge engineering.* Although this term has been used in association with the acquisition of knowledge for expert systems, we suggest a more encompassing definition to include the variety of functions that we see emerging as a broader engineering discipline.

8.4. New Imperatives of the CIM Paradigm

Knowledge and the management of intelligence have supplanted systems science as the primary domains of activity in the CIM era. The professional workstation has replaced the simple electronic gauge; versatility in the creation of new products and processes is the primary driver; the ability to generalize and abstract from experience in order to create new products is the required skill. The consequences of these changes are illustrated in Table 8.2 , which compares one factory's performance before and after conversion from Numerical Control to FMS.

The CIM paradigm dictates a new set of management imperatives:

- **Build small, cohesive teams**. Very small groups of highly skilled generalists show a remarkable propensity to succeed.

- **Manage process improvement, not just output**. FMS technology fundamentally alters the economics of production by drastically reducing variable labor costs. When these costs are low, little can be gained by reducing them further. The

		NC	FMS	Ratio
Factory output (to make comparison clear, these have been held constant)	Types of parts produced per month	543	543	1.0
	Number of pieces produced per month	11,120	11,120	1.0
Space	Floor space required (m^2)	16,500	6,600	2.5
Equipment	CNC machine tools	66	38	1.7
	General purpose machine tools	24	5	4.8
	Total machines	90	43	2.1
People – 3 shifts	Operators	170	36	4.7
	Distribution, production control workers	25	3	8.3
	Total people	195	39	5.6
Average processing time per part (days; includes queue time)	Machining time	35	3	12.
	Unit assembly	14	7	2.0
	Final assembly	42	20	2.1
	Total processing time	91	30	3.0

Table 8.2 Performance of one factory before and after FMS [19]

challenge is to develop and manage physical and intellectual assets, not the production of goods. Choosing projects that develop intellectual and physical assets is more important than monitoring the costs of day-to-day operations. Old-fashioned, sweat-of-the-brow manufacturing effort is now less important than system design and team organization.

- **Broaden the role of engineering management to include manufacturing.** The use of small, technologically proficient teams to design, run, and improve FMS operations signals a shift in focus from managing people to managing knowledge, from controlling variable costs to managing fixed costs, and from production planning to project selection. This shift gives engineering the line responsibilities that have long been the province of manufacturing.

- **Treat manufacturing as a service**. In an untended FMS environment, all of the tools and software programs required to make a part have to be created before the first unit is produced. Although the same is true of typical parts and assembly operations, the difference in an FMS is that there is no capability for in-the-line, people-intensive adjustments. As a result, competitive success increasingly depends on management's ability to anticipate and respond quickly to changing market needs. With FMS technology, even a small, specialized operation can accommodate shifts in demand. Manufacturing now responds much like a professional service industry, customizing its offerings to the preferences of special market segments.

9

The New World of Work
– Intelligence, Volatility, Dynamism

We have come full circle. The new manufacturing environment looks remarkably similar to the world of Maudslay. Expert workers, with high discretion, conduct a wide range of activities as needed. Feasible product variety and customization in Maudsley's era and today are essentially infinite. Figure 9.1 traces the progression of work from high to low discretion and back again, and from increasing mechanization in an essentially static world, to increasing intelligence in a more volatile dynamic world. Even the number of workers in a state-of-the-art factory now and in 1810 are similar – a few dozen skilled experts. (Figure 9.2)

The first three epochs increasingly focused on machines and what could be done with them in a static environment. Under Taylor the role of direct labor became to adapt to the machines and the environment – to be yet another machine.

The glue that makes a collection of machines a manufacturing system is people processing information. The need for integration, and the intelligence needed to make machines function, were the focus of the three epochs in the last half of the 20th century. In them, we see a reversal of the trends of mechanization of repetitive work: increasing

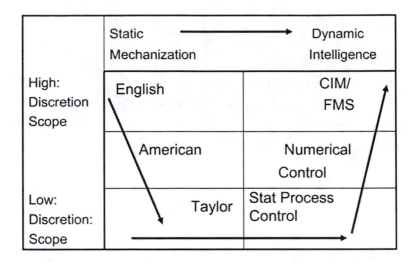

Fig. 9.1 Changing nature of work across six epochs

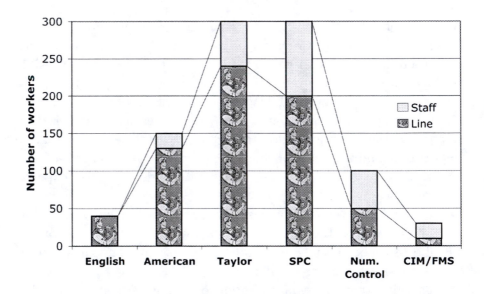

Fig. 9.2 Evolution of work force size and mix across six epochs

versatility and intelligence; substitution of intelligence for capital; and economies of scope. Machines have increasingly been seen as extensions of the mind; the goal of the engineering workstation is to enhance the cognitive capabilities of the human being.

Of course, many other characteristics of manufacturing have evolved monotonically. Taylor's concept of executing tightly specified repetitive procedures has not vanished – it has instead been offloaded from workers to machines. The level of automation has increased, and the number of workers per machine has dropped by two orders of magnitude (Figure 9.3). We can say that as process control shifts from art to science, the fraction of staff – those workers who manipulate knowledge and information – increases, the execution of procedure is increasingly embodied into machines, and performance (rework, tolerances, and productivity) improves. The holy grail of a manufacturing science, begun in the early 1800s and carried on with religious fervor by Taylor in the early 1900s with the dawning of the twenty-first century, is finally within grasp. [7]

We have seen that in each of the six epochs of process control, what has changed is not just technological demands, but also the organization of manufacturing to meet those demands, and the nature of work. We have also seen that the fundamental shifts involved in the transition from one epoch to the next are intellectual shifts. Each epoch brought a new class of core problems, and therefore a new way of posing and solving problems. The roles of workers, the nature of work, and the key focus of technology all had to shift. To the extent that a firm competes by acquiring, developing, and managing know-how, these intellectual shifts become technological imperatives. One can argue that the shifts themselves are socially determined, and that technology is a social product. Nevertheless, insofar as we live in a competitive world, we must, once one of these shifts has occurred, adapt to the technological imperatives imposed by it.

We have seen, in detail for each of the six epochs, what the changes have meant for manufacturing management in the firearms industry. Is it possible to generalize these findings to other industries? As long as there are structural similarities in the manufacturing process technologies – metal fabrication, for instance – we would venture that the broad thrust of the argument holds. There is a paradigmatic shift to a more dynamic, information intensive world, centered on the develop-

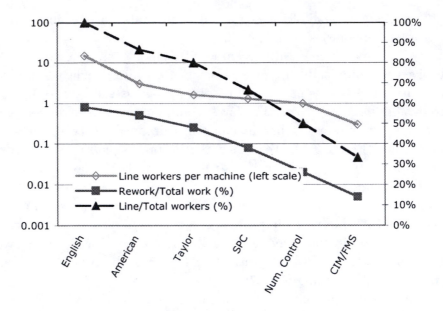

Fig. 9.3 Evolution of work and quality

ment of intellectual assets. Managing these intellectual assets, specifically attending to the man-machine cooperative system, is the new
challenge. [22]

References

[1] H. G. J. Aitken, *Taylorism at Watertown Arsenal; scientific management in action, 1908-1915*, Harvard University Press, Cambridge, Mass., 1960.

[2] Anonymous, "A Day At The Armory Of 'Colt's Patent Fire-Arms Manufacturing Company'," *United States Magazine*, vol. IV, no. 3, March, 1857, http://www.connhistory.org/wwsevis_reading.htm.

[3] Anonymous, *Gun Manufacture*, McGraw Hill, New York, 1942.

[4] Anonymous, *Modern Small Arms*, Penton Publishing Co., Cleveland, Ohio, 1942.

[5] Anonymous, "Set of Inspection Gauges for US Rifle Model 1841," In: *Yankee Enterprise: The Rise of the American System of Manufactures*, Mayr, O. and Post, R. C., Smithsonian Institution Press, Washington, D.C., pp. 78, 1981.

[6] Charles Babbage, *On the Economy of Machinery and Manufacture*, 1832, http://onlinebooks.library.upenn.edu/webbin/gutbook/lookup?num=4238.

[7] Roger E. Bohn, "From Art to Science in Manufacturing: An Extension of Jaikumar's Study in the Evolution of Process Control," *Foundations and Trends™ in Technology, Information and Operations Management*, 2005.

[8] James Bright, *Automation and Management*, HBS Press, 1958.

[9] Brown and Sharpe Manufacturing Co., "Smithsonian Institution Negative 81-202," In: *Yankee Enterprise: The Rise of the American System of Manufactures*, Mayr, O. and Post, R. C., Smithsonian Institution Press, Washington, D.C., 1981.

[10] Denis Diderot, "Arquebusier," In: *Encyclopédie, ou dictionnaire raisonné des sciences, des arts et des métiers*, Diderot, Denis, Paris, 1751–1772.

[11] Charles H. Fitch, *Report on the Manufacture of Fire-Arms*, Government Printing Office, 1882.

[12] Constance McLaughlin Green, *Eli Whitney and the Birth of American Technology*, Little Brown, Boston, 1956.

[13] C. T. Haven and F. A. Belden, *A History of the Colt Revolver*, William Morrow and Company, New York, 1940.

[14] Wilma Pitchford Hays, *Eli Whitney: Founder of Modern Industry*, Franklin Watts, Inc., New York, 1965.

[15] *Hearings before the Special Committee of the House of Representatives to Investigate the Taylor and Other Systems of Shop Management*, 62nd Congress, Washington, D.C., 1912.

[16] Charles Holtzapffel, *Turning and Mechanical Manipulation*, London, Pub. for the author, by Holtzapffel & co., 1843-50, 1847.

[17] F.R. Hutton, *Report on Machine-Tools and Wood-Working Machines*, Government Printing Office, Washington, DC, 1885.

[18] Imperial-Royal Lombard Institute for Science, Letters and Art, "Notes on the Arms Industry in Gardone in the Trompia Valley," In: *Beretta: La dinastia industriale più antica al mondo: the world's oldest industrial dynasty*, Morin, Marco and Held, R., Acquafresca editrice, Chiasso, Switzerland, 1980.

[19] Ramchandran Jaikumar, "Postindustrial Manufacturing," *Harvard Business Review*, vol. 64, no. 6, pp. 69-76, 1986.

[20] Ramchandran Jaikumar, *Contingent Control of Synchronous Lines: A Theory of JIT*, HBS, working paper, May, 1988.

[21] Ramchandran Jaikumar, *From Filing and Fitting to Flexible Manufacturing: a Study in the Evolution of Process Control*, Harvard Business School, working paper, 1988.

[22] Ramchandran Jaikumar and Roger E. Bohn, "The Development of Intelligent Systems for Industrial Use: A Conceptual Framework," In: *Research on Technological Innovation, Management, and Policy*, Rosenbloom, Richard S., JAI Press, London and Greenwich, Connecticut, vol. 3, pp. 169-211, 1986.

[23] Ramchandran Jaikumar and Roger E. Bohn, "A Dynamic Approach to Operations Management: an Alternative to Static Optimization," *International Journal of Production Economics*, vol. 27, no. 3, pp. 265-282, 1992.

[24] Robert G. Landers, A. Galip Ulsoy, and Richard J. Furness, "Process Monitoring and Control of Machining Operations," In: *The Mechanical Systems Design Handbook*, CRC Press, 2002.

[25] D. S. Landes, *The Unbound Prometheus*, Cambridge University Press, Cambridge, England, 1969.

[26] O. Mayr and R. C. Post, *Yankee Enterprise: The Rise of the American System of Manufactures*, Smithsonian Institution Press, Washington, D.C., 1981.

[27] Marco Morin and R. Held, *Beretta: La dinastia industriale più antica al mondo: the world's oldest industrial dynasty*, Acquafresca editrice, Chiasso, Switzerland, 1980.

[28] James Nasmyth, *Autobiography of James Nasmyth*, London, 1883.

[29] David F. Noble, *Forces of Production: A Social History of Industrial Automation*, Alfred A. Knoppf, 1984.

[30] J.W. Roe, *English and American Tool Builders*, McGraw-Hill, New York, 1926.

[31] Gordon V. Shirley and Ramchandran Jaikumar, "Turing Machines and Gutenberg Technologies: The Post-Industrial Marriage," *ASME Manufacturing Review*, vol. 1, no. 1, pp. 36-43, 1988.

[32] M. R. Smith, *Harper's Ferry Armory and the New Technology*, Cornell University Press, Ithaca, NY, 1985.

[33] Paul Uselding, "Measuring Techniques and Manufacturing Practice," In: *Yankee enterprise, the rise of the American system of manufactures: a symposium*, Mayr, Otto and Post, Robert C., Smithsonian Institution Press, Washington DC, 1981.

[34] Abbott Usher, *A History of Mechanical Inventions*, Harvard University Press, 1954.

[35] Johann Christoff Weigel, *Verschiedene Stucke Fur Buchsemacher*, Originally in *Plusiers Models des plus nouvelles manieres qui sont en usage en l'Art d'Arquebuzerie*, Paris, 1660. Reproduced from Diderot's *Encyclopedia*, "Arquebusier," courtesy of the Kress Library of Business and Economics, Harvard Business School, 1702.

[36] Whitworth, *Presidential Address, Institution of Mechanical Engineers*, pp. 125, 1856.

[37] Robert S. Woodbury, *Studies in the History of Machine Tools*, MIT Press, 1972.

Author's Biography

Born June 18, 1944, in Madras, India, Ramchandran "Jai" Jaikumar graduated from the Indian Institute of Technology (Madras) in 1967. He spent much of each year during college climbing in the Himalayas, and a first ascent earned him the right to name Mt. Leela. He once described his college academic performance as staying one step ahead of failing until the climbing season ended, then earning top grades for the rest of the year. In one climbing accident after a successful ascent, he and his climbing partner fell hundreds of meters. His partner was killed and Jaikumar broke several bones. Without a tent, he had to hike through the night in hopes of finding shelter, and was ultimately rescued by a Nepalese peasant who carried him on her back to the nearest town.

After earning an MS in industrial engineering from Oklahoma State University, Jaikumar worked as an operations researcher at Sara Lee. Subsequently he entered the PhD program in Decision Sciences at the Wharton School, where he and his advisor, Marshall Fisher, developed new techniques in very large scale integer programming for vehicle routing. These became the basis of a company they founded.

On graduating from Wharton in 1980 Prof. Jaikumar joined the Technology and Operations Management group at the Harvard Business School, where he spent the rest of his career, ultimately becoming

Daewoo Professor of Business Administration. Although he published further papers on optimization, most of his research at Harvard took a very different approach to the issues of modern manufacturing and technology. His research on the integration of computers and manufacturing led him to develop a "minimalist" architecture for manufacturing that emphasizes elimination of process disruptions and simplification of control over elaborate calculations.

Prof. Jaikumar earned numerous honors, including the Frederick Winslow Taylor Medal, from the American Society of Mechanical Engineers, and the Grosvenor Plowman Prize, awarded by the National Council for Physical Distribution Management. He was a three-time finalist and double winner of the Franz Edelman Prize for management science practice.

Prof. Jaikumar often mixed consulting with research, and collaborated on many of his projects. He would invite friends to join him in some obscure location for hiking or bird watching while discussing research. A frequent visitor to Europe, he was well known for mixing travel with pursuit of gourmet food. "Jai's rule of restaurant selection," namely that "Satisfaction equals quality divided by time," led him to McDonald's and three-star restaurants, but little in between. In one typical gourmet quest, after spending the day in meetings in Northern Italy he drove for hours to reach a particular restaurant. When his party arrived at midnight, the restaurant was empty of customers, but the staff had stayed to prepare them a fresh meal. Another hobby was the history of technology, and Prof. Jaikumar's house was home to artifacts such as early clocks, climbing gear from a 1924 attempt on Mount Everest, an Edison lab notebook, and a first edition of Diderot's *Encyclopedia*.

Prof. Jaikumar taught the required MBA course for much of his career at Harvard and was known for his energy and encyclopedic memory of students. During semesters when he was teaching he used the Concorde to make brief consulting trips to Europe. On one occasion he was late for a crucial connecting flight and was stopped in Switzerland for speeding. He explained his situation to the policeman, who agreed to send him on his way but with two tickets: one for the speeding he had just done, the other for the speeding he was about to do. Despite further misadventures he made it back to Boston in time for his class.

Jaikumar made two other first mountain ascents, in Greenland and Bolivia. He named the Greenland peak 'Mount Minarjnik,' a contraction of the names of his wife, Mrinalini Mani, and sons Nikhil and Arjun. He died of a heart attack on February 10, 1998 while climbing a volcano near Quito, Ecuador.

Foundations and Trends® in
Technology, Information and Operations Management
Vol 1, No 2 (2005) 121-202
© 2005 R.E. Bohn

From Art to Science in Manufacturing: The Evolution of Technological Knowledge

Roger E. Bohn

*Graduate School of International Relations
and Pacific Studies
University of California, San Diego
La Jolla, CA 92093-0519,
USA*

Rbohn@ucsd.edu

Abstract

Making goods evolved over several centuries from craft production to complex and highly automated manufacturing processes. A companion paper by R. Jaikumar documents the transformation of firearms manufacture through six distinct epochs, each accompanied by radical changes in the nature of work. These shifts were enabled by corresponding changes in technological knowledge. This paper models knowledge about manufacturing methods as a directed graph of cause–effect relationships. Increasing knowledge corresponds to more numerous variables (nodes) and relationships (arcs). The more dense the graph, the more variables can be monitored and controlled, with greater precision. This enables higher production speeds, tighter tolerances, and higher quality.

Changes in knowledge from epoch to epoch tend to follow consistent patterns. More is learned about key classes of phenomena, including measurement methods, feedback control methods, and disturbances. As knowledge increases, control becomes more formal, and operator discretion is reduced or shifted to other types of activity. Increasing knowledge and control are two dimensions of a shift from art towards science.

Evolution from art to science is not monotonic. The knowledge graphs of new processes are riddled with holes; dozens of new variables must be identified, understood, and controlled. Frederick Taylor pioneered three key methods of developing causal knowledge in such situations: reductionism, using systems of quantitative equations to express knowledge, and learning by systematic experimentation.

Using causal networks to formally model knowledge appears to also fit other kinds of technology. But even as vital aspects of manufacturing verge on "full science," other technological activities will remain nearer to art, as for them complete knowledge is unapproachable.

1

Introduction

Since the first Industrial Revolution, technology has steadily transformed living standards and daily life. The aggregate effects of new technology – rising productivity and improving product performance – are visible effects of from new knowledge of "how to do things." But what is the nature of this knowledge, and how does it evolve over time? This paper investigates long-term technological change and the evolution of enabling knowledge through the lens of a single industry over more than 200 years.

Changes in technological knowledge are usually observed indirectly, as changes in methods or performance. Performance that improves by more than can be explained by measured inputs is taken as evidence of changes in the stock of knowledge. Implicitly this assumes a causal chain approximately as follows: learning activities create new knowledge that allows the firm to implement superior designs and methods that improve local physical performance such as machine speed and material consumption, which ultimately causes better high level performance (Figure 1.1). But generally, the middle variables in this chain are not observed directly.

Our focus is on the intermediate steps of this chain – new knowledge, superior methods, and improved performance at workstations –

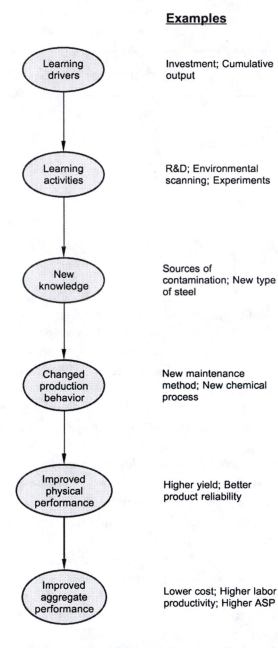

Examples

Investment; Cumulative output

R&D; Environmental scanning; Experiments

Sources of contamination; New type of steel

New maintenance method; New chemical process

Higher yield; Better product reliability

Lower cost; Higher labor productivity; Higher ASP

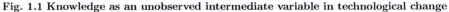

Fig. 1.1 Knowledge as an unobserved intermediate variable in technological change

that cause improved aggregate performance. Changes in production methods are explored in the companion paper, *From Filing and Fitting to Flexible Manufacturing: The Evolution of Process Control* by R. Jaikumar [15]. Here we explicitly examine the new knowledge that made possible these changes.

Our case study centers on the manufacturing methods of a single company over 500 years. The company, Beretta, has remained in family hands and has made firearms since its founding in 1492, when firearms were manufactured as a small-scale craft with only hand tools. Jaikumar identified six distinct epochs of manufacturing, characterized by different conceptions of work, different key problems, and different organizations (Table 1.1).

Each epoch constituted an intellectual watershed in how manufacturing and its key activities were viewed. Each required introducing a new system of manufacture. Machines, the nature of work, and factory organization all had to change in concert. Within Beretta, each of these epochal shifts took about ten years to assimilate.

A longitudinal study of a single industry is an excellent test-bed to examine technological change over a long period. In Jaikumar's study, the fundamental product concept changed little from the 16th to the late 20th century: a chemical explosion propels a small metal object through a hollow metal cylinder at high speed. With such product stability, changes in manufacturing stand out even more.

The central problem in manufacturing over the entire period was to increase process control, for once society moved beyond making unique items by hand predictability, consistency, and speed were achieved by progressively tightening control. Each new epoch revolved around solving a new process control challenge, generally reducing a novel class of variation. To accomplish this required major, often unexpected, shifts in many aspects of manufacturing (Table 1.2). The nature and organization of work changed, use and sophistication of machines increased, and, most important for our purposes, manufacturing control shifted, all requiring changes in knowledge.

We will describe shifts in technology using the metaphor of transformation *from art to science*. Jaikumar observed that "The holy grail of a manufacturing science begun in the early 1800s and carried

Epoch	Approx date
0) The **Craft System** (circa 1500)	1500
1) The invention of machine tools and the **English System** of Manufacture	1800
2) Special purpose machine tools and interchangeability of components in the **American System** of Manufacture	1830
3) Scientific Management and the engineering of work in the **Taylor System**	1900
4) **Statistical process control** (SPC) in an increasingly dynamic manufacturing environment	1950
5) Information processing and the era of **Numerical Control** (NC)	1965
6) Flexible manufacturing and **Computer-Integrated Manufacturing** (CIM/FMS)	1985

Table 1.1 Manufacturing epochs [15]

on with religious fervor by Taylor in early 1900s is, with the dawning of the twenty-first century, finally within grasp."[1] But precisely what does this mean? Is such evolution inevitable? Is it universal, or limited to manufacturing?

As late as the early 18th century, making firearms still relied entirely workers' expertise. Documented or standardized methods were non-existent.

> Production involved the master, the model, and a set of calipers. If there were drawings, they indicated only rough proportions and functions of components. Masters and millwrights, being keenly aware of the function of the product, oriented their work towards proper fit and intended functionality. Fit among components was important, and the master was the arbiter of fit. Apprentices learned from masters the craft of using tools. Control was a developed skill situated in the eyes and hands of the millwright.

---

[1] Unattributed quotations are from [15].

		English System	American System	Taylor System	Statistical Process Control	Numerical Control (NC)	Computer Integrated Manufacturing
Size Trends	Introduced at Beretta (world)	1810 (1800)	1860 (1830)	1928 (1900)	1950 (1930)	1976 (1960)	1987
	# of People (Min. Scale)	40	150	300	300	100	30
	Number of Machines	3	50	150	150	50	30
	Productivity Increase*	4:1	3:1	3:1	3:2	3:1	3:1
	Number of Products	Infinite	3	10	15	100	Infinite
Nature of work	Standards for Work	Absolute product	Relative product	Work standards	Process standards	Functional standards	Technology standards
	Work Ethos	"Perfection"	"Satisfice"	"Reproduce"	"Monitor"	"Control"	"Develop"
	Worker Skills Required	Mechanical craft	Repetitive	Repetitive	Diagnostic	Experimental	Learn/generalize/abstract
	Control of Work	Inspection of work	Tight supervision of work	Loose of work/tight of contingencies	Loose supervision of contingencies	No supervision of work	No supervision of work
	Organizational Change	Break-up of guilds	Staff-line separation	Functional specialization	Problem-solving teams	Cellular control	Product/Process/Program
	Staff/Line Ratio	0:40	20:130	60:240	100:200	50:50	20:10
	Line Workers per Machine	15	3	1.6	1.3	1	0.3
Technology Keys	Process Focus	Accuracy	Precision: Repeatability (of machines)	Precision: Reproducibility (of processes)	Precision: Stability (over time)	Adaptability	Versatility
	Focus of Control	Product functionality	Product conformance	Process conformance	Process capability	Product/process integration	Process intelligence
	Instrument of Control	Micrometer	Go/No-Go gauges	Stop watch	Control chart	Electronic gauges	Professional workstations
	Rework**	.8	.5	.25	.08	.02	.005

*Over previous epoch **As fraction of total work

Table 1.2 Summary of epochal changes [15]

> Inasmuch as adaptive skills are really contingent responses to a wide variety of work conditions, procedures cannot readily be transferred. Critical knowledge was mainly tacit, and a journeyman had to learn by observing the master's idiosyncratic behaviors. The master, who could solve the most difficult of problems, fashioned each product such that quality was inherent in its fit, finish, and functionality. [15, Section 2]

This description corresponds to technology as an art. Learning was by apprenticeship; quality was achieved by rework; progress occurred slowly by trial and error; techniques and knowledge were idiosyncratic.

In contrast, in the most advanced flexible manufacturing systems of the late 20th century people are normally absent from the production area, and machines execute complex contingent procedures under computer control. Operators manipulate symbols on workstations, and use scientific methods of observation, experimentation, and data analysis. Alternative production methods can be precisely described, tested, and embodied in software. Methods and general knowledge can be transferred to other locations, machines, and products with little effort and no face-to-face communication. This is manufacturing as a science. Manufacturing changed profoundly over the two century transition from art to science, with performance improvements on some dimensions of two orders of magnitude or more (Figure 1.2).

Transitions from art toward science can be seen in many technologies. Early aviation, literally a "seat of the pants" technology early in its development, today includes the Global Hawk aircraft, which can take off, cross the Pacific, and land without human intervention. In contrast, although product development technology has progressed tremendously, it still has remains in many ways more like art than a science.

Although we are concerned here with a relatively small industry that has not been leading edge since the mid-19th century, the evolution of knowledge and the transition from art to science are still critical in all high-tech industries, and influence many contemporary issues such as offshoring, automation, and outsourcing. These activities require transfers of knowledge and information across organizational and firm

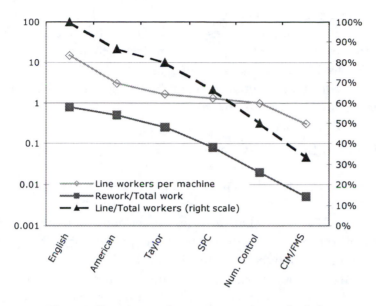

Fig. 1.2 Changing performance over six epochs [15]

boundaries. We will see that the difficulty of such transfers depends on the detailed structure of knowledge. [18]

In Section 1.1 we consider different ways of classifying technology along a spectrum from art to science. Section 1.2 presents a formal model of technological knowledge that supports precise descriptions of changes in knowledge when learning occurs. Prior research is presented in Section 1.3. The case study evidence is presented in Section 2 and Section 3.

In Section 2 we examine the first three epochs of manufacturing (approximately the 19th century), during which workers' discretion and insight were progressively reduced, culminating in Taylor's extreme division of labor and separation of intellectual work from line operations. We will see that the de-skilling of workers in the Taylor System rested on an unprecedented level of technological knowledge, developed by Taylor himself using several seminal concepts.

In Section 3 we examine the development of knowledge over the last three epochs, in which workers increasingly became problem solvers and knowledge creators, effectively reversing Taylor's de-skilling paradigm. We also examine the integration of formal science with practical engineering. Finally, we consider what happens when novel and immature physical processes are substituted for mature ones. Even when the core physical process is entirely changed, considerable knowledge from old processes is still relevant.

In the concluding section we examine broad patterns of change in manufacturing over the centuries.

1.1. Art and Science in Technology

The metaphor of art and science in human endeavor is long established and widely used. Military treatises speak of the "art and science of war" as in a 1745 book that provides "a short introduction to the art of fortification, containing draughts and explanations of the principal works in military architecture, and the machines and utensils necessary either in attacks or defenses: also a military dictionary ... explaining all the technical terms in the science of war" [3]. Sometimes a clear distinction is made between "art" and "science," as in the title of an American book on surveying circa 1802: *Art without science, or, The art of surveying: unshackled with the terms and science of mathematics, designed for farmers' boys* [33]. The two are not as clearly differentiated in a 1671 title, *An introduction to the art of logick: composed for ... [those who do not speak Latin but] desire to be instructed in this liberal science* [28].

In modern usage art and science are generally viewed as the extremes of a spectrum. "Art" conveys the sense of a master craftsman using informal and tacit knowledge, "science" that of an engineer who uses mathematical equations to program computerized machines. Furthermore, the outcome from a craftsman is not as predictable or consistent as that from an engineer. The sense (whether legitimate or not) is that amateurs and low-volume production are at the artistic end, professionals and high-volume production at the scientific end (e.g., a home cook versus a packaged goods bakery). Most technical and managerial activities are perceived to require a mix, and progress in understanding a field to correspond to a gradual shift from "mostly art" to "mostly science."

The range of methods in human endeavor can be examined along many dimensions. Particularly useful for characterizing technology are how work is done, quality of the results that are achieved, and how well the technology is understood (Table 1.3). Any of these dimensions could be used to specify some measure of "art or science" and we might expect that all move toward "science" as a given technology advances.

The activity dimension describes how actions are carried out, whether according to rigid procedures or idiosyncratically. Procedure refers to specifying activities in advance and reducing them to complete and explicit rules that must be followed exactly. [16] We observe this in a lights-out factory, in which every intentional action results from explicitly stated computer instructions executed properly by microprocessors. Human discretion characterized the pre-manufacturing world of expert craftsmen who used rudimentary hand tools, informal judgment, and individualized methods without formal guidelines. But it is simplistic to equate degree of procedure with the extent of automation, which would imply that activities done by machines are fully rigid and those performed by people cannot be. Much of the emphasis during the Taylor epoch was on applying rigid procedure ("one best way") to people, and in many factories today this continues to be the goal.

We can also characterize art-versus-science by the nature of the *knowledge* about a given technology. If nothing is known production is impossible; if everything is fully understood, we can call it completely science-based. We will analyze how knowledge moves from one extreme towards the other, through intermediate gradations. Among the criteria used to describe knowledge qualitatively the most common is probably the degree of explicitness – whether knowledge is tacit or codified. Polanyi pointed out that much knowledge cannot be written down, even when it is critically needed in order for a technological system to function properly.[2] [30]

Codified knowledge refers to knowledge that is transmittable in formal, symbolic language, whereas tacit knowledge is hard to articulate and is acquired through experience ... Tacit and codified

---

[2] The literature on this topic is vast; Balconi's analysis of tacit knowledge in modern manufacturing is similar in spirit to that in this paper [5].

	Embryonic technology: "Art"	Ideal technology: "Science"
How *activities* are executed	Zero procedure; idiosyncratic	Fully specified procedure
What *results* are achieved	Each one different, mostly poor	Consistent and excellent
Characteristics of *knowledge*:		
How knowledge specified	Tacit	Codified
What knowledge about	Purely know-how	Also know-why
Extent of knowledge	Minimal; can distinguish good from bad results, but little more	Complete

Table 1.3 Dimensions of production technology on an art-science spectrum

knowledge exist along a spectrum, not as mutually exclusive categories ... For some knowledge, especially [sic] in medical practice, the difference between tacit and codified is temporal: much codified knowledge in medicine today was tacit in the past. [14]

Knowledge that tells *what* to do but does not explain *why* things happen is also incomplete. For example, it is inadequate to debug problems.[3] Finally, we can examine the quality of the *results* achieved by a process. A perfect technology should always deliver perfect results, especially in conformance quality. At the other extreme, pure art would never produce the same thing twice, and much of what is produced is expected to be unusable.[4]

Movement along the dimensions of action, knowledge, and results (Table 1.3) tends to occur in concert, in part because the extent of available knowledge constrains procedures. For example, all desired

---

[3] Know-how and know-why are often referred to as *procedural knowledge* and *causal knowledge*. See the discussion of [23] later. An additional category is *declarative knowledge*.

[4] Many other ways of classifying knowledge are used. For example, the distinction between collective and individual knowledge is important for designing knowledge management systems. [2]

actions to be performed by a numerically controlled (NC) machine tool must be specified in detail in computer programs, which are highly formal procedures. Writing effective programs requires that knowledge be extensive and explicit. When these conditions are not met, procedures can still be specified but will not work well.

Each step of a process can be summarized by two measures, the amount known about it and the degree of procedure used to execute it (Figure 1.3). If these are consistent, points plotted on a graph will be near the diagonal, and over time a process step will normally move up and to the right. If knowledge is inadequate for the degree of procedure used, the plotted point will be above the diagonal and the step will not operate well. Conversely, if a process is below the diagonal, it could have been done in a more formal way, presumably reducing cost and improving consistency.

The increasingly formal execution of manufacturing from epoch to epoch is detailed in [15], corresponding to upward movement in Figure 1.3. Implicitly this requires greater knowledge. We address this next.

1.2. A Model of Technological Knowledge

New methods, if they are to be superior to their predecessors, must be based on new knowledge (Figure 1.1). To understand how technological knowledge changes and grows over time requires a disaggregated model, detailed enough to compare two knowledge states. Notwithstanding the substantial body of research on innovation and technology, specific knowledge is little analyzed in the technology management literature.

> Recent studies of engineers, scientists and technicians have brought to light the social and political aspects of work ... [but] as a whole they overemphasize the importance of political actions and social networks and underestimate the importance of formal, often technical, knowledge in the carrying out of tasks. Formal knowledge looms in the background in nearly every study of technical workers. [4]

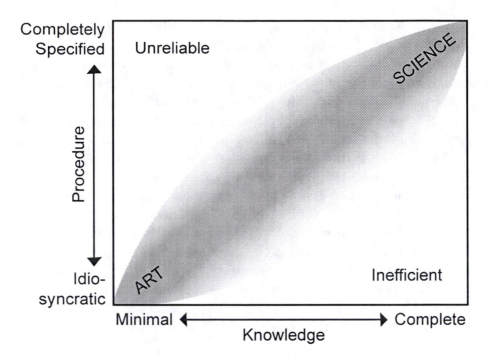

Fig. 1.3 Level of knowledge versus degree of procedure (adapted from [11])

To do our analysis we therefore develop and exploit a model of technological knowledge. It starts with the following observation:[5]

> The core of technological knowledge is knowledge about causality in human-engineered systems.

Designing, building, or operating a technological system, whether a firearm or a factory, requires an understanding of the causal relationships among actions, events, and outcomes. Only with such knowledge can desired outcomes be achieved, and undesirable ones debugged.

Causality can be modeled formally using causal networks, directed graphs whose nodes are variables. Directed arcs between the nodes

---

[5] This theory of technological knowledge was developed jointly with R. Jaikumar. Previous work includes [9].

show causal relationships.[6] [29] Variables can be physical properties of an object, logical values, or information. Useful variables for a metal part, for example, might include its composition, shape, mass, hardness, and perhaps color. For a machine tool, they include control settings such as speed and feed, actual behavior such as cutting depth and vibration, and many elements of its design. These variables are linked in a dense network of causal relationships, and the state of knowledge at any moment can be summarized by depicting the causal network as it was understood (implicitly or explicitly) at that time. This known causal network expands as technological knowledge develops.

Relationships among variables can also be described by mathematical functions, in particular by systems of nonparametric simultaneous equations. Any such system can be summarized by a causal network. The simplest relationship is two variables A and B that cause a third variable C, C = f(A, B). (Left side of Figure 1.4.) The properties and arguments of the function f are known only to a limited extent. Better knowledge about the technology corresponds to better understanding of the causal network's topology and of the specifics of the function f.

Genealogical terminology is used to express causal relationships. A *parent* causes a *child* if there is a direct link from parent to child. Parents often have many children, and children usually have many parents. *Descendants* are all nodes that can be reached by forward chaining from a variable and, equivalently, whose values may be affected by it. *Ancestors* include parents, grandparents, and so forth: any variable of which the child is a descendant. In Figure 1.4, E and F are both descendants of A, B, C, and D; E is a child only of D and F is a child of both B and D. Cycles are possible; one variable can be both ancestor and descendant of another. Such relationships create feedback, such as would occur if there were a directed link from E to A. A *causal path* from X to Y is a directional sequence of ancestors of Y, each variable having the previous one as its only parent in the chain. A ⇒

---

[6] Pearl's formal definition is "A causal structure of a set of variables V is a directed acyclic graph (DAG) in which each node corresponds to a distinct element of V, and each link represents direct functional relationship among the corresponding variables." (page 43) Note that this definition specifies *Acyclic* Graphs, which cannot have feedback. But feedback loops are central to process control and are central to any theory of modern technological knowledge. Therefore, we will allow cyclic graphs.

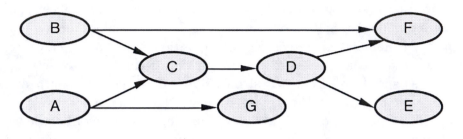

Fig. 1.4 Simple causal network

C $\Rightarrow$ D $\Rightarrow$ F is a causal chain from A to F; B $\Rightarrow$ F is also a chain. *Cousins* are variables with at least one common ancestor, but no causal path from one to another (such as D and G). Cousins are statistically correlated, but no causal relationship exists among them.

One value of causal networks is that they quickly identify how a variable can be altered; all and only its ancestors can affect it. Because of this property, any unexplained change in a variable reveals the existence of previously unknown parents. Causal networks also facilitate various kinds of counter-factual reasoning (difficult or impossible with standard statistical models), such as predicting how a system will behave under novel operating rules. [29] They thus not only represent knowledge abstractly, but also constitute useful knowledge in themselves.

It is often useful to select a small number of important variables that summarize the important results of a system. These are referred to as *outcome* variables for that system. In manufacturing, typical outcome variables are production rates, costs, and properties of the final product. These variables are chosen based on criteria from outside the system: *the ultimate goals of a causal system are selected exogenously*. Typical goals of a production system might include cost minimization, high conformance quality, and high output.

Causal networks reveal how the outcomes are determined by their ancestors. Each ancestor, in turn, has its own network of ancestors. The important input variables for one process include the outcome variables for upstream processes and suppliers, including the properties of machines or materials passed from one to the other. In this way, causal paths can be traced back through industrial supply chains.

Causal networks for production processes are extremely complex, but not all variables are equally important. The status and behavior of a process or sub-process can generally be summarized by a few important intermediate variables. Good intermediate variables are often "choke points" in the causal network – many ancestral variables determine their levels, and they in turn exert multiple effects. They can include machine control settings, process behavior, and physical properties of products. Simply learning the identities of key variables is useful, and often requires considerable effort.

As Jaikumar showed, fabricating accurate parts by machining was a key activity throughout the history of firearms manufacture. Figure 1.5 shows a highly simplified causal network for machining. The most important variable, metal removal, is at the center. The shape and location of the metal removed from a workpiece are functions of the motion of the cutting tool relative to the workpiece surface, the cutting tool characteristics, and the composition and orientation of the workpiece before cutting begins. Behavior of the cutting tool is determined by the various processes that created or affect it, e.g. those related to the machine power train and to tool maintenance. A variety of machine adjustments enable workers to influence results, for example by changing the cutting depth. Almost without exception, adjustments are based on some form of feedback control. For example, in the pre-numerical control epochs an experienced machinist used sound, the shapes of chips from the workpiece, and other indicators to determine whether and how to adjust cutting. Higher order feedback loops (not shown in the figure) are used to diagnose systemic problems, and many small feedback loops embedded in subsystems' control variables such as motor speed.

The causal network in Figure 1.5 emphasizes desired process variables and relationships. But what makes manufacturing especially challenging are undesired disturbances. An operator can set the *intended* behavior of a machine, but not the actual behavior. Disturbances arise both from outside the system, such as defective raw materials, and as side effects such as vibration and contamination. We will see that no

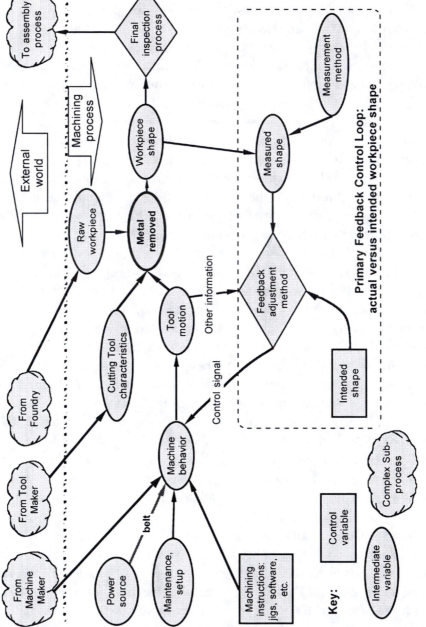

Fig. 1.5 Stylized causal knowledge graph for machining

complex process can be completely understood, much less measured in real time, so detecting disturbances, uncovering their sources, and devising counter-measures are never-ending stories. The causal knowledge graph includes whatever is known about disturbances and their effects.

This paper limits the domain of inquiry to "hard engineering." We will analyze only knowledge about *human-designed systems intended to accomplish tangible physical tasks.* This excludes analysis of, among other things, worker motivation, strategic goal-setting, and political interactions among people and organizations. The virtue of limiting the domain so sharply is that objective truth exists, even if it can never be fully known.

Axiom: *The true causal network exists, is complete, and is deterministic.*

Pearl stated this as follows:

> We view the task of causal modeling as an induction game that scientists play against Nature. Nature possesses stable causal mechanisms that, on a detailed level of description, are deterministic functional relationships between variables, some of which are unobservable. [29, p 43]

Applying this specifically to machining:

> The following statement is the basis of the Deterministic Theory: "Automatic machine tools obey cause and effect relationships that are within our ability to understand and control and there is nothing random or probabilistic about their behavior" (Dr. John Loxham). Typically, the term random implies that the causes of the errors are not understood and cannot be eradicated ... The reality is that these errors are apparently nonrepeatable errors that the design engineers have decided to quantify statistically rather than completely understand. [20]

The evolution from art toward science occurs through identifying, in ever greater detail and breadth, "Nature's stable causal mechanisms."

In the following sections we analyze the knowledge about them that emerged in each epoch.

There is an important distinction between the *true causal network* and what is believed at a particular time in a specific organization. The true causal network exists and is deterministic, but it is never fully known. The belief network, in contrast, is never perfect or complete. For example, metal removal by a cutting tool can be described by a system of algebraic equations first crudely set down only circa 1900. Yet the relationships summarized by such equations were always active. For clarity, the known version of the true causal network will be referred to as the causal *knowledge graph*. This also sidesteps the problem that the organizational learning literature uses the term "knowledge network" to mean something entirely different.

1.2.1. Stages of Knowledge

The causal knowledge graph gives the overall structure of knowledge about a technology. The structure of the graph is only a partial description of what is known. The degree of knowledge about specific variables and relationships (nodes and arcs in the graph) shifts qualitatively as more is learned, passing through a series of stages. We use an extension of the framework from [16] and [9].

Knowledge about individual variables can be classified into six stages (Table 1.4). Initially, many of the variables in a process are not even recognized (Stage 0). Other variables in the same process might be almost completely understood and controlled. In between, knowledge about a variable has several possible degrees.

Similarly, two variables might be recognized as somehow related (for example they may be statistically correlated), but the nature of their relationship not known. With effort, more might become known about how one variable causes the other (Table 1.5).

Each node and each arc in a causal graph has its own stage of knowledge. Some combinations of stages, however, are impossible. For example, a variable cannot be adjustable unless at least one of its parents is adjustable and the magnitude of the relationship between them is known.

Stage	Name	Description	Comment
0	Unknown	Complete ignorance: the existence of X is not known	Effects of X perceived as pure noise
1	Recognized	The existence of X is known, but magnitude is only known qualitatively. Even ordinal measure may not exist.	X is an exogenous disturbance
2	Measurable	X can be measured on a cardinal scale, through a repeatable measurement process	
3	Adjustable	The mean level of X can be altered at will but the actual level has high variation	X is endogenous to the process
4	Capable	Control of the variance: Enough is known to reduce the variance of X to a fraction of its uncontrolled level	X can be used as a control or outcome variable for the process
5	Perfectly Understood	Complete knowledge: X can be held at a target level under all conditions.	Stage 5 knowledge is unreachable; it can only be approached asymptotically.

Table 1.4 Stages of knowledge about control of an individual variable

1.3. Other views of technological knowledge

The most thorough analysis of specific technological knowledge is Vincenti's work on aeronautics. [41] *What Engineers Know and How They Know It* contains five detailed case studies of how specific aeronautical problems were solved, including the design of airfoils, design of propellers, and design and production of flush riveting. The cases cover the development of theoretical design tools, a series of empirical experiments to reveal the effects of design choices in the absence of adequate theory, and the case of riveting, in which dimensional tolerances and design of tools played key roles.

Vincenti classifies the knowledge developed by engineers in the case studies into six categories:

Stage	Name	Description
0	Ignorance	No awareness that X and Y might be related. The true effects X on Y are perceived as a random disturbance
1	Correlation	Aware X and Y are related but not nature of causality (ancestor, descendant, or cousin)
2	Direction	Direction of causality known (X a cause of Y, not a descendant or cousin)
3	Magnitude	Know the partial derivative of Y with respect to X (or shape of the partial function, for highly nonlinear relationships)
4	Scientific model	Scientific model: Have a scientifically based theory giving functional form and coefficients of relationship between X and Y
5	Complete	Complete knowledge. Stage 5 knowledge is unreachable; it can only be approached asymptotically

Table 1.5 Stages of knowledge about the relationship between two variables (True relationship: X an ancestor of Y)

- **Fundamental design concepts**: The operational principles and normal configuration of working devices.

- **Criteria and specifications**: Specific criteria and quantitative targets for key intermediate variables. Examples include load per rivet, dimensional tolerances, and "stick force per g of gravity."

- **Quantitative data**: Usually from experiments, and represented by tables or graphs.

- **Practical considerations**: Knowledge about issues that have little formal role, but nonetheless influence how something should be designed (e.g., the capabilities of specific machines).

- **Theoretical tools**: A broad category that includes intellectual concepts such as feedback, mathematical tools such as Fourier transforms, and theories based on scientific principles such as heat transfer.

- **Design instrumentalities**: Knowledge about how to design, such as structured design procedures, ways of thinking, and judgmental skills.[7]

Although he does not use the metaphor of art versus science, Vincenti is conscious of the progression of design knowledge and procedures from crude to exact, or as he puts it, from "infancy to maturity." For example, he summarizes the development of airfoils as follows.

Finally we can observe – somewhat roughly – a progression of development in airfoil technology, which I take to comprise both explicit knowledge and methods for design. The first decades of the century saw the technology in what can be called its infancy. No realistically useful theory existed, and empirical knowledge was meager and uncodified. Design was almost exclusively by simple cut-and-try; that is, by sketching an airfoil and trying it out. No other way was possible. Today, airfoil technology has reached maturity. Using relatively complete (though not yet finished) theories, supported by sophisticated experimental techniques and accurate semitheoretical correlations of data, engineers design airfoils to specific requirements with a minimum of uncertainty. Little cut-and-try is needed by a skilled professional. Between the phases of infancy and maturity lay a half-century of growth. In this period theory provided qualitative guidance and increasing partial results, but wind-tunnel data were vital. Design was an uncertain and changing combination of theoretical thinking and calculation and cut-and-try empiricism ... Perhaps we could call this decade [of most rapid change, from late 1930s to early 1940s] the adolescence of airfoil technology, when rational behavior was on the increase but offbeat things could still occur. Whether or not we push the metaphor that far, we can at least see a progression of development through phases of infancy, growth, and

---

[7] This list has been extended by Bailey and Gainsburg's study of building design, which added construction feasibility, organization of work, and engineering politics [4].

maturity, with a characteristic relationship of knowledge and design in each phase.[8] [41; p 50]

Vincenti is concerned with design, not manufacturing. Nonetheless, if we substitute "art" or "craft" for infancy, and "nearly perfect science" for maturity, his formulation of the transformation of technology from art to science is consistent with what we will describe for firearms manufacture. We will return to Vincenti's classification of technological knowledge, which encompasses more issues but is less precise than the one used here, in the last section.

---

[8] Additional work on the evolution of knowledge using Vincenti's framework includes [12] and [42].

2

Evolution of Knowledge in a World of Increasing Mechanization

Machine tools, invented circa 1800, brought mechanical power and control to metal shaping. During the first three epochs of manufacturing, from 1800 to the early 20th century, the precision of these machines was progressively increased, mainly by mechanical means that constrained the behavior of machines and workers. The key developments of this period emphasized knowledge about different portions of the machining process (see Figure 1.5).

Little formal knowledge about any portion of the machining process existed prior to 1800. Quantitative measurement of parts not yet existing, the goal was to make each new firearm as similar as possible to the shop's working model. Even the conformance of finished parts to the model was judged idiosyncratically, by eye and caliper. Beyond this little can be said. Plates from Didier's *Encyclopedia* illustrate the range of hand tools available and undoubtedly there was qualitative knowledge (both verbal and tacit) about when and how to use them to achieve desired results.

2.1. English System

Different epochs emphasized the development of knowledge about different subsystems of processes. The state of technological knowledge in the English System is little documented, but we can infer general properties of the knowledge from what was achieved during that epoch. Technological breakthroughs revolved around three subsystems: the machine, specification of intended outcomes, and measurement of actual outcomes (Table 2.1).

Maudsley's achievement of highly accurate parts measurement using micrometers was accompanied by the invention of the engineering drawing. Accurate measurement and an absolute goal provided by the engineering drawing enabled a distinction between "better" and "worse" parts, which otherwise would have been judged merely "different" as in the Craft epoch. Taken together, the micrometer and the engineering drawing supported the creation of a basic feedback loop: keep removing material until a part is of the dimension specified in the drawings as measured by a micrometer. [15, Section 3]

Woodbury described Maudsley's other key contribution, the general purpose machine tool with highly precise lead screws for accurately cutting parts with a minimum of trial and error, in the four key elements: ample power and drive train sufficient to effect its delivery; adequate rigidity under the stress of cutting ferrous metal; precision in construction greater than the precision of the parts to be produced; and adjustability to accommodate flexibility in the parts. [44, pp 96–97] At a minimum, enough was thus known to design and build iron machines with these properties.

2.2. American System

The American System introduced new concepts of ideal outcomes based on tolerances and precision as well as accuracy. The corresponding new measurement method was the use of go/no-go gauges.

"Accuracy in this system, which might be as close as a thirty-second or sixty-fourth of an inch, was ensured by an elaborate system of patterns, guides, templates, gauges, and filing jigs." The use of these

Key Invention	Portion of process (Figure 1.5)	Significance
Machine tool w. lead screw	Machine	Accuracy in cutting
Engineering drawing (projective geometry)	Specify target shape	Ability to state desired goal and measure actual outcome enable feedback control for finer accuracy than can be delivered by the machine tool
Micrometer, standard plane	Measurement method	

Table 2.1 Key knowledge contributions of the English System

geometric devices to constrain the motion of cutting tools required the development of causal knowledge about linkages from jigs to final parts (Figure 2.1).

Colt and others developed, in parallel with knowledge about making firearms, the knowledge needed to design and build machine tools for specific purposes. Workers independent of those employed in the manufacture of firearms "built, maintained, set up, and improved machines." Specialized machine tool companies emerged to sell these machines abroad to furnish entire firearms factories.

Implicit in the emergence of these companies is another fundamental innovation of this epoch: separation of organizational knowledge by causal module. A machine tool designer does not need to know what parts are to be fabricated, only how to construct a machine capable of cutting along precise trajectories. The parts maker need not understand the nuances of how the machine works, only a limited range of adjustment methods. Information is transmitted from one to the other through the jigs. This separation of toolmakers' from tool users' knowledge is vital to the success of capital equipment industries.

What conditions support this separation of users and suppliers? There are two key conditions, one physical, the other having to do with knowledge.

First, the technology itself must have a modular causal network, that is, the total causal network must be separable into two subnetworks with much denser connections within than between them. The comparatively few connections between the subnetworks must be almost

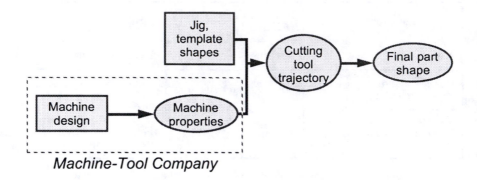

Fig. 2.1 Causal network from jigs to part shapes

entirely in a single direction. Such a network structure is observed, for example, with geographically separated suppliers and customers between which there is a one-way flow of intermediate product. Causal paths tying the firms together pass through these intermediate products.

Second, knowledge about the causal relationships that join the subnetworks must be sufficiently complete to enable the modularity to be exploited. The key relationships that link the subnetworks must be well understood and their variables be known and measurable.

If both conditions are met, each subnetwork can be controlled by its own organization (department or firm) and the two joined by an arms length relationship. In Figure 2.1 , a cutting tool's trajectory is a function of only a limited number of machine tool properties. Knowledge about the causal linkages among these properties was sufficient in the American System to make separate machine tool companies feasible.

2.3. Taylor System

Their extensive research on the "hard" technology of machining would render the impact of Taylor and his team on the transition from art to science fundamental, even in the absence of their more well known work at the Watertown Arsenal on worker procedures and standardized methods for each job. Conducted in secret for more than 20 years, the research was finally presented, in 1906, to an overflow audience of 3,000 at a gathering of the American Society of Mechanical Engineers. [35]

Key Invention	Portion of process (Figure 1.5)	Significance
Elaborate jigs and fixtures	Control method	Precision and flexibility possible
Go/no-go gauges	Measurement method	Simple way to estimate precision
General purpose machine tools	Machine	Separation of machine knowledge from product knowledge; organizational specialization

Table 2.2 Key knowledge contributions of the American System

As in the other epochal shifts Taylor did not so much add to the established body of knowledge in its own terms, as shift the nature of the knowledge sought. His fundamental contributions to technological knowledge were several (see Table 2.3).

- Taylor's reductionist approach to systems analysis divided parts production into linked subsystems, each carefully analyzed in isolation to arrive at a formally specified "best" process. He studied not only parts machining, but also indirect supporting activities.

- Taylor moved from qualitative and ordinal relationships among variables to systems of equations with numerical coefficients that could be solved quantitatively.

- Finally, he employed a much superior learning method, namely a large number of carefully controlled empirical experiments, to develop knowledge systematically.

- These three contributions enabled Taylor's team to make specific discoveries about better manufacturing methods, perhaps most important their discovery of high-speed steel.

Each of Taylor's contributions constitutes a move from art towards science. The scientific knowledge he developed was a prerequisite for the development of standardized work procedures – his "one best way" – for which he is more famous. In Taylor's view, the best

Key Invention	Portion of process (Figure 1.5)	Significance
Concepts of repeatable process, separable subsystem	Ancillary subsystems (e.g., power)	Allows separation and improvement of staff activities, reductionism in analysis
Simultaneous equation models to describe complex causal relationships	Metal cutting	Represents knowledge in explicit and easily manipulated form
High speed (heat treated) steel; other specifics of cutting methods	Cutting tool	Huge improvement in feasible cutting speeds, costs
Carefully controlled experiments; four-step learning process	Learning method (not shown)	Facilitates discovery of quantitative causal knowledge for any repeatable process

Table 2.3 Key knowledge contributions of the Taylor System

way could be determined only after the behavior of each subsystem was understood and had been quantified. Thus, for each subsystem, he moved towards science along the knowledge axis in advance of corresponding movement along the procedural axis. We consider these advances in turn.

2.3.1. Reductionist Approach to Manufacturing Systems

Taylor's insight was that production encompassed a host of distinct processes that could be analyzed and improved independently of the larger system they comprised. The sharpening of a tool, in his view, could be managed and optimized independently of the purpose for which the tool was to be used. As with the separation of capital equipment from firearms manufacture in the American System, this is feasible if and only if there is causal knowledge modularity. Taylor further realized that separation, analysis, and improvement could be applied to auxiliary processes such as accounting and maintenance as well as to materials processing.

Taylor applied this approach to all activities that had a significant effect on the overall rate of production, for example, the power transmission system (pink areas in Figure 1.5). The electrical motors of Taylor's day were large and expensive, so a few central motors powered dozens of machine tools by means of a network of moving belts. [15, Figure 5.1]

> Inasmuch as the speed of operators was largely determined by the speed of the machines as driven from a central location by belts, pulleys, and shafts, Taylor considered the standardization and control of these systems at their optimal level of efficiency essential. To this end he established the activities of belt maintenance and adjustment as a separate job and prescribed methods for scientifically determining correct belt tensions. [15, Section 5]
>
> A great deal of the old belting was replaced with new and in some cases heavier belting. This made it possible to run machines at higher speeds and with greater power, so that full advantage could be taken of the cutting powers of high-speed steel, and also prepared the way for Barth's later standardization of cutting speeds and feeds. By the end of April 1910 the belt-maintenance system was in full operation and belt failures during working hours had been practically eliminated. [1]

Taylor studied and optimized the causal subnetwork that determined belt breakage and other belt-related influences on production rates (Figure 2.2). Belt failures had persisted despite limiting speeds. By standardizing and optimizing the belt maintenance system (B in Figure 2.2), the tradeoff between speed and belt reliability was substantially shifted outward, enabling faster speeds (A and D) while reducing the incidence of breaks (C). Since total production is the product of cutting rate and operating time, productivity improved substantially.

Taylor developed for the first time detailed knowledge and corresponding procedures for many other subsystems.

- Standardization of ancillary equipment (e.g., sockets, screws)
- Storeroom handling of in-process materials
- Tool maintenance (including tool room procedures and equipment)

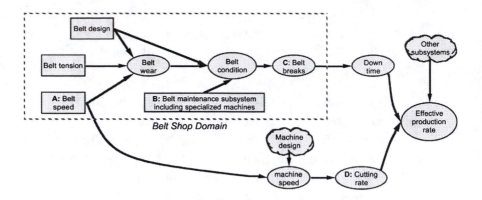

Fig. 2.2 Belt-related causal knowledge graph

- Cutting speeds (discussed below)
- Tool design and fabrication, especially the metallurgy of new high-speed steels

For each subsystem, analyzing and prescribing behavior required the development of knowledge about at least three parts of its causal network.

- The *key outcome variables* that describe the results of the subprocess
- The *ancestral causal network* used to deduce what caused the outcomes including the identities of and relationships among *key intermediate variables*
- The *best levels* of key control variables not only for specific cases but also for ranges of operating requirements.

Just to establish which variables are important is no small task. In his 26-year investigation Taylor identified twelve groups of variables that affected optimal cutting speed (Table 2.4).[1]

---

[1] This list is from [36], with modern terminology provided by [26]. Cutting speed is a key outcome variable because it directly drives output and total factor productivity.

Variables that influence optimal cutting speed (from [35])	Magnitude of effect
• Quality (e.g., hardness) of the metal to be cut	100
• Depth of cut	1.36
• Work piece's feed per revolution	3.5
• Elasticity of the work or tool	1.15
• Shape or contour as well as clearance and rake angles of the cutting edge of the tool	6
• Tool material (e.g., chemical composition and heat treatment)	7
• Use of a coolant such as water	1.4
• Tool life before regrinding	1.2
• Lip and clearance angles of the tool	1.023
• Force exerted on the tool by the cut	Not given
• Diameter of the work piece	Not given
• Maximum power, torque, and tool feeding force available on the lathe	Not given

Table 2.4 Taylor's list of key variables related to cutting speed

In a seminal lecture and paper, Taylor presented these variables in terms of their effects on optimal cutting speed. The numbers in the last column are his estimates of the sensitivity of cutting speed to each variable. For example, the most potent decision variable is tool material, reflecting the importance of Taylor's discovery of high-speed steel and the way machining procedures had to change to take advantage of it. [15]

2.3.2. Expressing Causal Knowledge as Systems of Equations

Organizing variables as in Table 2.4 yields a simple causal structure in the manner of the shallow tree depicted in Figure 2.3. Taylor and his team recognized, however, that behavior was driven by systems of nonlinear equations (although they did not use that terminology). They eventually expressed the relationships as equations such as:

$$V_{20} = \frac{\text{Constant}\left(1 - \dfrac{8}{7(32r)^2}\right)}{\left\{f\exp\left(\dfrac{2}{5} + \dfrac{2.12}{5+3r}\right)\right\}\left\{\dfrac{48d}{32r}\exp\left(\dfrac{2}{5} + 0.06\sqrt{32r} + \dfrac{0.8(32r)}{6(32r) = 48d}\right)\right\}}$$

where V_{20} = cutting speed that leads to a 20 minute tool life, in feet per minute
r = tool nose radius, in inches
f = feed per revolution, in inches
d = depth of cut, in inches [26].

These equations, derived empirically by fitting curves to experimental data, were too complex to solve, but the team was able to embody approximations of the most important into specialized slide rules (see Figure 2.4).[2] Each slide rule is an analog computer corresponding to a specific system of multivariate equations, and some were specific to a single machine. With them the values of the respective variables could be solved for, given values of enough of the other variables.[3]

Multiple slide rules with common variables were used to solve for multiple outcome variables. Cutting conditions, for example, were used by one slide rule to determine how much power the machine tool would require, by another to determine how much stress would be placed on the spur gears (Figure 2.5), and by a third to determine how long the cutting operation would take.

Some of Taylor's results are still used today. A summary relationship known as the Taylor equation, for example, is used to trade off cutting speed versus tool life, both of which have direct economic effects.

---

[2] Taylor does not discuss how the curves were fit to data and he does not try to justify the functional forms he used. This was before the use of statistical analysis for experimental data and his data tables suggest heavy use of judgment. [35 exhibits]
[3] Solution methods are described in elaborate detail in [7].

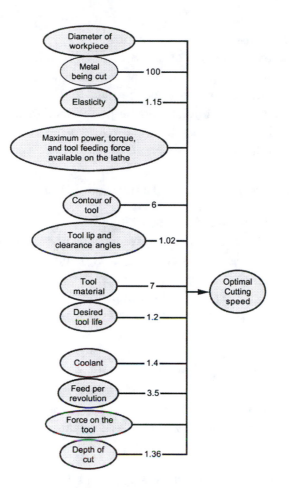

Fig. 2.3 Possible simplistic knowledge graph for cutting influences

$$VT^n = C$$

where V = cutting speed in feet per minute,
T = cutting time to produce a standard amount of tool wear,
n is an empirical constant for the material being cut,
and C an empirical constant for other cutting conditions such as tool design and material.

There are still no general predictive models for n or C, but engineering handbooks have tables of n for different metals and C can be estimated experimentally for a given situation. Figure 2.6 shows the corresponding causal knowledge graph.

The existence and use of these formulas, slide rules, and corresponding systems of equations show the extent of causal knowledge developed by Taylor. Not only did he identify the important *variables* that govern how machining should be done for high production rates, he also claimed that he understood the *relationships* among the variables well enough to derive *normative rules* for the best way to machine.

2.3.3. Systematic Learning by Experimentation

The third fundamental way in which Taylor and his team moved machining knowledge from art towards science was through a systematic learning methodology. This comprised two major innovations, (1) a procedure for learning about any subprocess, and (2) massive systematic experimentation to estimate the quantitative relationships in causal networks. Used with numerous subsystems, as for the infamous experiments on shoveling by "Schmidt," the most elaborate applications of these methods were in the areas of metallurgy of cutting tools and formulating cutting equations.

With the benefit of a century of hindsight we can see that Taylor developed a more-or-less-repeatable procedure for learning about physical causality. He organized the analysis and prescription of behavior for each subsystem into four steps.

(1) Identify the *key outcome variables* that describe the results of the subprocess. To make these variables operationally useful required establishing standard definitions and measurement methods. Taylor spent several years establishing the best way to measure tool wear, for example.

(2) Determine the *ancestral causal graph* for these outcomes including the identities of, and to the extent possible, important relationships among, key intermediate variables.

(3) Given this knowledge, determine the *best levels* of key control variables not just for specific cases but for ranges of operating requirements and conditions.

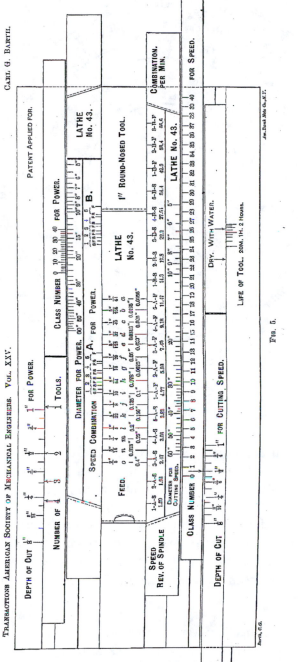

Fig. 2.4 Slide rule for key cutting variable relationships [7]

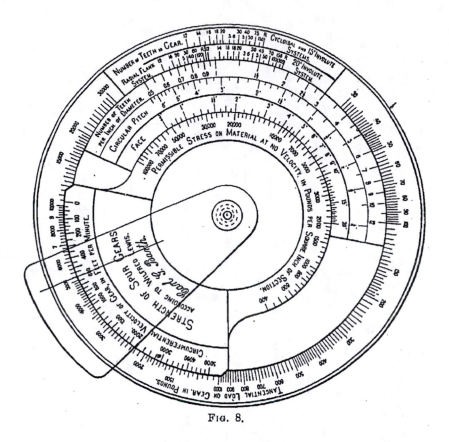

Fig. 2.5 Slide rule for side calculation [7]

(4) Establish *standard procedures* that make it easy for workers
    to use the best methods. This step essentially translated
    increased knowledge into formal procedures.

As important as his overall procedure was Taylor's use of massive
numbers of controlled experiments. He identified key variables and
relationships (steps 1 and 2) from experimental evidence. Taylor summarized his team's decades of experimentation on tooling and cutting
speed as follows.

Experiments in this field were carried on, with occasional interrup-
tion, through a period of about 26 years, in the course of which

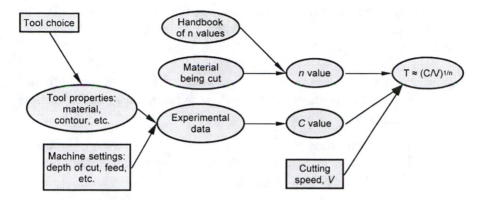

Fig. 2.6 Knowledge graph corresponding to Taylor equation $VT^n = C$

ten different experimental machines were especially fitted up to do this work. Between 30,000 and 50,000 experiments were carefully recorded, and many other experiments were made, of which no record was kept. In studying these laws [sic] more than 800,000 pounds of steel and iron was cut up into chips with the experimental tools, and it is estimated that from \$150,000 to \$200,000 was spent in the investigation. [36]

Taylor devoted many pages of his exposition to experimental methodology, both successes and problems, as in the following passage.

[W]e had made one set of experiments after another as we successively found the errors due to our earlier standards, and realized and remedied the defects in our apparatus and methods; and we have now arrived at the interesting though rather humiliating conclusion that with our present knowledge of methods and apparatus, it would be entirely practicable to obtain through four or five years of experimenting all of the information which we have spent 26 years in getting. [35, p 42]

Taylor also acknowledges "failure on our part from various causes to hold all of the variables constant except the one which was being systematically changed."

But Taylor reserved his most devastating critiques for academics and other perceived experts. His criticisms of previous research included the following: [35, p 40ff]

- That researchers assumed they knew which variables were and were not important, and ran their experiments such that these assumptions were never tested;

- That researchers conducted detailed investigations of complex and difficult-to-measure variables of no actual importance, in particular obsessive investigation of the pressure exerted on the cutting tool, which "calls for elaborate and expensive apparatus and is almost barren of [effect]";

- That researchers were also guilty of the converse; "several of those elements [variables] which are of the greatest importance have received no attention from experimenters" he complained, adding by way of example that "the effect of cooling the tool through pouring a heavy stream of water upon it, which results in a gain of 40 per cent in cutting speed, ... [has] been left entirely untouched by all experimenters";

- That researchers used "wrong or inadequate standards for measuring" dependent variables;

- That researchers changed multiple variables at once and in *ad-hoc* fashion.

Taylor's assessment of the best known previous research, conducted at the University of Illinois, is scathing. If his overflow audience hoped to be entertained as well as informed, they surely were not disappointed.

These experiments, from a scientific viewpoint, were so defective as to make it out of the question to deduce formulae, because no effort was made to keep the following variables uniform: (1) the shape of the ... tool varied from one experiment to another; (2) the quality of the tool steel varied; (3) the [heat] treatment of the tool varied; (4) the depth of the cut varied from that aimed at; (5) the cutting speed was not accurately determined at which each tool would do its maximum work throughout a given period of time;

and (6) ... it does not appear that any careful tests were made to determine whether [the raw unfinished workpieces being cut were] sufficiently uniform throughout in quality ... The same criticism, broadly speaking, applies to both the German and the University of Illinois experiments. [35, p 46]

Taylor's attention to detail (his biographers have commented on his obsessive personality) was vital to the success of his experiments and accounts for some of his major serendipitous discoveries. A modern description of Taylor's breakthrough development of high-speed steel portrays it as a premeditated and rational process, in marked contrast to Taylor's own account of his work.

Their investigation thus turned from the optimization of cutting conditions to the importance of heat treatment. Putting on one side conventional craft wisdom and the advice of academic metal-lurgy, Taylor and White conducted a series of tests in which tools were quenched from successively higher temperatures up to their melting points and then tempered over a range of temperatures. This work was made possible by use of the thermocouple which had not long been in use in industrial conditions. After each treatment, cutting tests were carried out on each tool steel ... Certain tungsten/chromium tool steels gave the best results ...

... The tools treated in this way were capable of machining steel at 30 [meters per minute] under Taylor's standard test conditions. This was nearly four times as fast as when using [the best previous] steels and six times the cutting speed for carbon steel tools. This was a remarkable breakthrough.

... High speed steels revolutionized metal cutting practice, vastly increasing the productivity of machine shops and requiring a complete revision of all aspects of machine tool construction. It was estimated that in the first few years, engineering production in the USA had been increased by $8 billion through the use of $20 million worth of high-speed steel. [40]

The tone of Taylor's description of this research is quite different. The breakthrough came when he attempted to demonstrate by running a trial in front of the foremen and superintendents of Bethlehem Steel his recent "discovery" that tools made from Midvale steel were the best. "In this test, however, the Midvale tools proved *worse* than those of *any* other make ... This result was rather humiliating to us." [35, p 51, emphasis added] Taylor's first reaction was to blame the workers who had made the sample tools, for heat treating them at too high a temperature. But this explanation was unproven and Taylor and his collaborator decided to characterize the exact effects of different temperatures. As expected, this revealed that tools were damaged by overheating to a temperature of around 1700 degrees F. But,

> to our great surprise, tools heated up to or above the high of 1725 degrees F. proved better than any of those heated to the best previous temperature ...; and from 1725 F. up to the [melting point], the higher they were heated, the higher the cutting speeds at which they would run.
>
> Thus, the discovery that phenomenal results could be obtained by heating tools close to the melting point, which was so completely revolutionary and directly the opposite of all previous heat treatment of tools, was the *indirect* result of an accurate scientific effort to investigate as to which brand of tool steel was [best]; *neither Mr. White nor the writer having the slightest idea that overheating would do anything except injure the tool* more and more the higher it was heated. [35 p 52, emphasis added]

Taylor's accounts of his research still elicit admiration. Although operating before the invention of statistical tools such as regression, design of experiments, and gradient search, Taylor clearly understood the importance of applying the scientific method. His sheer persistence and emphasis on careful empirical observation more than compensated for the inadequate statistical tools of his era.

2.3.4. Taylor's Legacy

Taylor wrought fundamental changes in the nature of work and in the procedural dimension of the evolution of manufacturing from art to science, in much the same way as did the English and American Systems of manufacture. [15] But the impact of the Taylor System on how technological knowledge is developed, partitioned, and expressed was even more revolutionary and fundamental. The concepts of learning through controlled experiments, of reductionism, and of expressing causal knowledge through systems of quantitative equations are still the bases of modern technology, and not just in manufacturing. Of course, Taylor's work was heavily influenced by its era; his methods had precedents in the natural sciences. But he harnessed their power and directed it at complex, real-world applications to manufacturing and process control.

Ironically, Taylor believed his innovations in factory management were more important than his work on machining and metallurgy. In his factories knowledge was not only developed independently for different activities, but was then used and maintained by staff specialists. In Taylor's shop, knowledge and execution were separated; workers were taught fixed methods for their jobs and only specialists were permitted to alter these procedures. But in the dynamic world the "one best way" changes frequently, and the necessary rates of problem solving and learning, which rely overwhelmingly on the intellectual abilities of workers, have increased dramatically. [17]

3

Knowledge in a Dynamic World

The treatment of knowledge changed fundamentally in the dynamic world that followed WW II. Problem solving and learning, which entailed the development of new knowledge, had to become organic to the production process. Finding a single optimum production method was replaced by change as the central concern of manufacturing.

> The first three epochs emphasized increasing mechanization in a world that was, at least ideally, static – doing the same tasks again and again, as efficiently as possible, at increasingly high volume. Discretion was progressively removed from workers, and knowledge about their tasks was subdivided and given to specialists, removing it from the shop floor ... In contrast, in the last three epochs, while the tools continued to become more mechanized, knowledge about the work was returned to workers and their discretion increased. The key goal shifted from efficiency at high volume to coping with a dynamic world of rapid changes such as high product variety and rapid product introduction. [15, Section 6]

But mechanization, in particular the development of increasingly autonomous machines, continued unabated. For a machine to operate autonomously a high level of knowledge is needed to guide responses to or forestall disruptions. Taylor's contributions to knowledge management discussed in the previous section thus continued to be vital, even as his approach to shop floor management was being turned on its head.

The Statistical Process Control (SPC) epoch coincided with a flowering of academic research on the science of metal cutting. Taylor's attempt to determine empirical formulas for factors that affect the rate of machining was extended, with the goal of raising effective machine speeds and productivity through a deeper understanding of the underlying science. This research was not primarily concerned with the SPC agenda of controlling variation.

The organization of this Section is not strictly chronological. We first examine the knowledge effects of the SPC epoch and the coinciding academic development of the "engineering science of machining." We then explore how numerical control initially foundered for want of sufficient knowledge. Finally, we consider what happens with fundamentally different manufacturing processes.

3.1. Statistical Process Control Epoch

The SPC epoch arrived at Beretta in the 1950s with the contract to manufacture the Garand M1 rifle. [15, Section 6] SPC shifted concern from average performance to variations in performance. To understand causes of variation requires detailed knowledge about a process and its real-world operation. Beretta's newly formed quality control department "was responsible for quantitatively measuring the natural variability of every machine and the degree of fidelity of every tool, verifying tool conformity to design, and identifying possible causes of systematic error."

Because so many variables can disturb a process, the complexity of causal networks for variation is an order of magnitude higher than for ideal operation. SPC thus drove the development of much more detailed causal knowledge, with a strong emphasis on the actual behavior of processes and machines on the factory floor.

This reorientation was accompanied by a complementary shift from a static to a dynamic world view. Dynamic causal models, in which sources and consequences of changes are explicitly monitored over time are vital to SPC. Each variable becomes a time series. Dynamic behavior such as the rate at which variables change had to go from being recognized to being measured (via control charts) to being adjustable. To eliminate adjustments between setups, for example, the rate of drift of key variables had to be constrained. But because dynamic behavior in this period was still not technically capable, processes escaped from control and interventions continued to be necessary.[1]

"Soft" innovations, such as control charts, were a hallmark of this period. The genius of the control chart is that it enabled operators, in a pre-computer era, to track dynamic variables and filter out real shifts from normal stochastic variation. Beretta's quality control department employed a variety of even more sophisticated statistical techniques such as gauge R&R studies, which are still essential for physical measurement.[2]

These changes shifted the focus of manufacturing from *control* to *learning*. "The application of SPC provided one way by which errors could, over time, be observed, better understood, and eventually solved. Manufacturing's evolution from an art to a science now included a systematic way of learning by doing." They also directed attention away from the *product* to the *process*. SPC effectively democratized and replicated Taylor's innovations in systematic learning about processes, even as his de-skilling of line workers was being reversed. Modern versions of SPC, such as Total Quality Management and Six Sigma, have institutionalized systematic learning, and moved it from the factory floor into general management.

---

[1] Tolerances were much tighter for the Garand rifle than previously (roughly .001 inches or 25 microns). Yet, rejects on Beretta's frame line went down from 15% to 3%, and overall rework time went down from 25% to 8%. Thus process capability improved even though tolerances tightened, suggesting that effective process variability was reduced by two orders of magnitude (standard deviation by one order of magnitude).

[2] Gauge R&R studies deal with the problem that measurements are inherently imperfect, and variation in measurements can be confounded with variation in the processes being measured, leading to serious mistakes. Gauge R&R also quantifies measurement variance from different sources. Although it is actually a sophisticated ANOVA calculation, training material teaches it as a "cookbook" procedure, and it can be done with little statistical knowledge.

Beretta's introduction of synchronous lines both required and made easier an integrated view of production, involving analysis of *interactions* among variables in different parts of a process. The sequence of workstations that comprise a process could no longer be assumed to be independent. This necessitated a major shift in problem solving and learning from a focus on individual machine performance to a process orientation.[3] "Diagnosis and problem solving are now carried out by examining the workstation not in isolation, but as part of the entire system ... Synchronous lines forced an integrated view of the entire system of manufacture. Whereas the intellectual underpinnings of Taylorism were reductionism and specialization, that of SPC in a synchronous line was integration."

3.2. The Science of Cutting Metal

At roughly the same time that Beretta was introducing SPC, formal laboratory-based research into machining was being conducted by universities and company research labs. Much of this research emphasized machining-speed issues in the Taylorist tradition, over precision and quality which are central concerns of SPC. A distinguishing feature was the effort to develop models based on known scientific principles rather than just fit curves to empirical data.

> The basic characteristic of science-based modeling of machining is that it draws on the established natural sciences, and particularly the science of physics, to establish reliable predictive models. These are models that can then be used to carry out reliable engineering calculations of the expected behavior or characteristics of a machining process, independent of empirical information.
> Development of capability for science-based modeling of machining was quite dependent on the knowledge and understanding of machining developed by the [earlier] research on empirical modeling. A good example of such was the research done by the

---

[3] The impact of synchronous lines on knowledge modularity is a topic in itself. One factor is that with no inspection or delay between workstations, problems in a workstation propagate downstream without any chance to be removed. By the time a problem is finally observed at the end of the line, it could have originated anywhere upstream.

Key Invention	Portion of Process (Figure 1.5)	Significance
Control chart	Higher order feedback system for controlling process (not shown)	Attention focusing for problem solving and learning; leads to continuous improvement
Synchronous line	Multiple workstations	Forces integrated perspective; interactions easier to study
Science-based models (see below)	Workpiece-tool interface	Causal knowledge more general; integrate scientific knowledge from diverse sources

Table 3.1 Key knowledge contributions of the SPC epoch

Ernst–Merchant team ... in the period from 1936 to 1957, which culminated in the creation by Merchant of the basic science-based model of the machining process. [26]

Researchers found, for example, that the shear angle, the angle at which metal chips "peel away" from the face being machined, was key to predicting machining behavior. Shear angle being an important intermediate variable, it became a target for detailed causal modeling. "The ultimate goal of the above analysis leading to the shear angle relationships is to enable the estimation of all the relevant metal cutting quantities of interest, such as the forces, stresses, strains, strain rates, velocities, and energies *without actually measuring them.* For example ... knowing the shear stress of the metal and the cutting conditions, all of the above metal cutting quantities of interest can be calculated." [19, p 86, emphasis added] We can thus say that for the first time the knowledge graph incorporated "first principle" scientific models.

Among its major accomplishments this research:[4]

- Extended Taylor's empirical research to a range of additional operations (turning, milling, drilling) and issues (surface finish, costs, forces);

---

[4] Following is based primarily on [26].

- Established a qualitative understanding of what happens when a tool cuts. The research identified four basic processes: primary shear, secondary shear, fracture, and built-up edge formation. These correspond to four distinct causal models with only modest overlap; [26]

- Yielded further details of cutting tool design, including materials and geometries for different purposes;

- Originated theoretically based models of the forces at work in metal cutting (e.g., Figure 3.1);

- Contributed analytic models of heat and thermal effects in metal processing.

In addition to incorporating fundamental scientific models for the first time, this research was notable for its depth. More variables and more relationships were incorporated into knowledge graphs, reflecting the *fractal nature of causal knowledge*. The more closely a phenomenon is examined, the more complex it appears. The effects include:

- Individual variables are replaced by collections of more specific variables.

- When a variable is discovered to be important, its causes must be understood in turn.

- New relationships among variables are identified, so a causal knowledge subgraph that is initially tree-like becomes a more complex network.

- Engineered subsystems are created to control new key variables. These systems add complexity beyond that of the underlying physical process. Even a simple feedback loop requires its own new causal system with measurement methods, a calculation algorithm, and an adjustment method.

Cutting tool geometry provides an example of the intricacy of knowledge. The Taylor experiments discussed previously showed the importance of heat treatment, which we now know affects the grain structure of the tool. Elemental composition of the steel is also

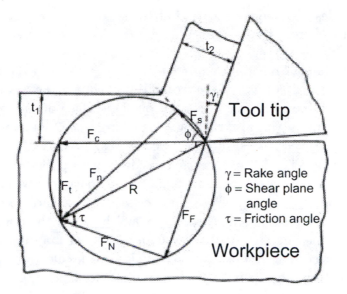

Fig. 3.1 Forces at work in chip cutting [25]

important. Tool geometry might seem more straightforward to describe, but six angles (and six corresponding dimensions) are required to begin to do so, and these six angles interact with more than 20 additional variables indicated by the underlined phrases.

Single-Point Cutting Tool Geometry. [A figure, not included here, shows] the location of [six] angles of interest on a single-point cutting tool. The most significant angle is the *cutting-edge angle*, which directly affects the shear angle in the chip formation process, and therefore greatly influences tool force, power requirements, and temperature of the tool/workpiece interface. The larger the positive value of the cutting-edge angle, the lower the force, but the greater the load on the cutting tool. For machining higher-strength materials, negative rake angles are used. *Back rake* usually controls the direction of chip flow and is of less importance than the side rake. Zero back rake makes the [chip] spiral more tightly, whereas a positive back rake stretches the spiral into a longer helix. *Side rake angle* controls the thickness of the tool behind the cutting edge. A thick tool associated with a small rake angle provides <u>maximum strength</u>, but the small angle produces higher cutting forces than a larger angle; the large angle requires less <u>motor horsepower</u>.

The *end relief angle* provides <u>clearance</u> between the tool and the finished surface of the work. <u>Wear</u> reduces the angle. If the angle is too small, the tool <u>rubs on the surface</u> of the workpiece and <u>mars the finish</u>. If the angle is too large, the tool may <u>dig into the workpiece</u> and <u>chatter</u>, or show <u>weakness</u> and <u>fail</u> through <u>chipping</u>. The *side relief angle* provides <u>clearance</u> between the cut surface of the work and the flank of the tool. Tool wear reduces the <u>effective portion</u> of the angle closest to the workpiece. If this angle is too small, the cutter <u>rubs</u> and <u>heats</u>. If the angle is too large, the cutting edge is weak and the tool may dig into the workpiece. The *end cutting-edge angle* provides <u>clearance</u> between the cutter and the finished surface of the work. An angle too close to zero may cause chatter with heavy <u>feeds</u>, but for a smooth finish the angle on <u>light finishing cuts</u> should be small. [24, p 13–13]

Even six angles and six dimensions do not come close to fully describing an actual cutting tool's geometry. Moreover, how the tool is made can have a major effect on its performance.

The design of tools involves an immense variety of shapes and the full nomenclature and specifications are very complex ... The performance of cutting tools is very dependent on their precise shape. In most cases there are critical features or dimensions, *which must be accurately formed* for efficient cutting. These may be, for example, the clearance angles, the nose radius and its blending into the faces, or the sharpness of the cutting edge. The importance of precision in tool making, whether in the tool room of the user, or in the factory of the tool maker, cannot be over estimated. This is an area where *excellence in craftsmanship is still of great value.* [39, p 7, emphasis added]

In other words, even where the effects of *using* tool features can be predicted, the causal network for *making* good tools is not well understood, and manufacturing them is closer to the art end of the spectrum even today.

The development of formal models of machining based on first principles generated considerable excitement, but appears to have had only limited impact on practice. One reason might be the tendency of academics to choose research issues based on the next logical intellectual problem rather than examine the most serious problems being encountered in the field. Jaikumar and Bohn [17] argue that in a dynamic world the critical problems tend to arise from poorly understood disturbances in real world manufacturing environments. Because not enough is known about them to simulate them in a laboratory, they must be studied on the factory floor, as Beretta did using SPC.

In some domains, moreover, theoretically grounded models did not agree well with experimental results.[19, p 86] One reason is that conditions (such as forces and temperatures) during metal cutting are much more extreme than those encountered during mechanical testing, where the relevant properties of materials are measured. Moreover, fundamental disagreements about correct ways to model particular phenomena persist. It is unclear, for example, whether the physics of metal cutting are sufficiently constrained to even have unique mathematical solutions.

Analysis of learning methods in another steel products industry, wire-making, illuminates the relationships among theoretical models, factory experimentation, and performance. [23 and articles cited therein] In one study, 62 process improvement projects were analyzed according to how extensively they developed theory-based causal knowledge ("conceptual learning") and how extensively they tested proposed changes on the factory floor ("operational learning"). Surprisingly, neither approach was sufficient to improve performance. Only projects that were high on both scales led to actual improvements, and many projects had a negative effect on performance. These results suggest that scientific models of metal processing can be helpful, but by themselves do not provide sufficient knowledge of real-world causality.

3.3. NC and CIM/FMS Epochs

In order for a computer program to successfully control a cutting tool, sufficient knowledge is needed first to predict how a process will behave, second to write a recipe that will reliably achieve the required tolerances, and third to either avoid or respond to disruptions without manual intervention. As tolerances tightened and adaptability became important, more detailed knowledge of the causal network was needed. (Table 3.2) This knowledge was not available when NC tools were first built and used.[5]

> The early problems of NC technology were partially due to limited formal knowledge of the machining process. A lot of the knowledge possessed by operators, such as when to make "on the fly" adjustments, was tacit or at least not accessible to programmers [and was therefore not incorporated into the NC programs]. This limited understanding of variations in machinability, tool wear, and part material properties, together with inadequate control strategies for coping with these shortcomings, significantly constrained early implementations of NC technology. But with effort, over time more of the tacit knowledge implicit in operator skills became precise, explicit knowledge that was used to develop procedures capable of dealing with a variety of contingencies. [15, Section 7]

Early implementations of numerical control were thus based on *less* knowledge than was accessible to conventional machinists, yet simultaneously employed a *higher* degree of procedure. The resulting attempt to operate above the diagonal region in Figure 1.3 resulted in frequent disruptions and poor outcomes.

> In the SPC era and before, master mechanics working with general purpose machines usually accrued years of experience, during which they accumulated a wealth of idiosyncratic knowledge about how to perform in a wide variety of circumstances. They talked in terms

---

[5] The term NC here covers both Numerical Control and Computer Numerical Control.

Key Invention	Portion of Process (Figure 1.5)	Significance
Hardware and software for machine control	Control system	Versatility
Special purpose algorithms for signal processing, dynamic control, and other	Measurement; adjustment	Sophisticated feedback and control despite noise
Variety of hardware sensors	Measurements	Monitoring or regulation of many variables in real time

Table 3.2 Key knowledge in NC epoch

of a "feel" for the machine, the tools, and the parts they worked on. It was through this feel that they were capable of producing parts to exacting specifications. Watching them work, one had a sense that they recognized errors (e.g., vibration, chatter, structural deformation due to thermal forces) as they were happening and adapted their procedures to compensate for them. This, in engineering terminology, is an advanced form of adaptive control in an ambiguous environment. Such adaptive error recognition and compensation requires either ... the experiential and partly tacit knowledge of the skilled machinist, or alternately a high stage of formal knowledge approaching full scientific understanding of the machinery, sensor, and controller technology, as well as of the product, the process, and all their interactions. [15]

For example, a potential problem in most machining is "chatter," a forced vibration of the tool against the workpiece that damages the surface as well as the tool. It is "easily detected by an operator because of the loud, high-pitched noise it produces and the distinctive 'chatter marks' it leaves on the workpiece surface." [21] Once detected, an experienced operator can stop it and even rework the damaged surface on the fly. But for an NC machine tool to detect chatter requires electronically sensing and processing an appropriate signal, usually sound. Due to the background noise that accompanies machining, this represents a difficult signal-processing problem for computers, compared with the excellent signal processing ability of the human nervous system. Having detected chatter, the NC machine must decide how to end it and, if possible, execute another pass to repair the surface finish. In other words, operators' knowledge about how to detect and deal with chatter must be replaced by adequate formal knowledge and complex signal processing. For many years, available formal knowledge was inadequate to solve this problem. Instead, NC programmers modified programs for particular parts to reduce cutting speeds. This avoided the domain in which chatter is likely to occur, at the cost of reduced productivity. Clearly, the more knowledge about when chatter will occur, the less safety margin is needed.[6]

To operate an untended FMS (Flexible Manufacturing System) requires even more knowledge than is needed to operate an equivalent set of NC tools. An FMS "lacks the stand-alone NC machine's almost constant attention from a machine operator, who can compensate for small machine and operational errors by realigning parts in a fixture, tweaking cutting tools, visually inspecting parts between workstations, and so forth." In the absence of this constant attention, small problems at one workstation can accumulate, and the number of possible contingencies that must either be prevented (which requires detailed understanding of their causes) or otherwise dealt with is much larger for FMS than for NC machines.

Consider the problem of tool breakage. A nearby operator can quickly detect a break, stop the machine, visually inspect the part for damage, instruct the machine to change tools, and take other corrective action. Although an operator can explain this sequence to an NC programmer, to equip a machine to detect a break is exceedingly complex. It took years for machine-tool makers to develop sufficient knowledge to add tool breakage and chatter detection to machine tools. Even when it detects a break, what response should the machine make? To diagnose the type of break and choose the best from among a set of possible responses requires considerable knowledge.

---

[6] A complementary approach developed later was to redesign machine tool structures to reduce the conditions under which chatter would occur. This required considerable research in applying mathematical theories of feedback and vibration. It is a superior solution in that it allows the tool to actually run faster without chatter. [26]

Jaikumar discusses the difficulties of problem solving in an FMS. The reason operators had to become knowledge workers, rather than vendors developing the necessary knowledge and selling their machines at a premium, is that much of the requisite learning and problem solving must be done at a local level. At the highest levels of speed and precision, individual machines exhibit idiosyncratic quirks that must be identified and compensated. Moreover, each plant, production line, and part number has specific characteristics and requirements. Owing to the interactions among all these variables, the preponderance of problems tend to be novel and local, although over time general knowledge can be built up and incorporated into machines and operating methods.

Research on process monitoring and response continues using a variety of advanced techniques. [21] A fundamental obstacle to process monitoring is that the working region of a machine tool is an extremely messy environment contaminated by coolant, chips, vibration, noise, dirt, and such. This exemplifies the problem of side effects. Energy applied in any form creates many children, of which only a few are desired. But all of the children propagate through the causal network, potentially causing disturbances at many points. Side effects are central to the nature of manufacturing and we return to them in the concluding Section.

The other development of the CIM/FMS epoch, computer integrated manufacturing (CIM), required additional knowledge about how to predict the behavior of part designs and manufacturing processes.

An engineer working with a number of different parts geometries could create and test [using simulation] different alternatives, settle on a tentative design, and then examine the manufacturing impacts of each part. A host of manufacturing related computer programs could then be used to create the NC programs needed to machine the components and even graphically display the tool path of a metal cutting program on the screen. When satisfied with the design, the engineer could transfer the program to a machining center and have the components fabricated automatically. [15, Section 8]

These simulations, and other capabilities embedded in CIM tools, rely on extremely high levels of knowledge about many phenomena. Complex systems of mathematical equations or equivalent algorithms are needed to model interactions among large numbers of variables and the knowledge generated must be "reduced to practice," that is, embedded in the CIM system, by applying additional formal knowledge of yet another kind. CIM software is never perfect; gaps become apparent as new manufacturing methods, product designs, and materials are introduced. Considerable improvement also occurs over time in the number of phenomena that can be incorporated, accuracy of the models, and speed of computational algorithms.

3.4. New Physical Processes

Firearms manufacture at Beretta makes a particularly good longitudinal case study of manufacturing in part because the core technology – the dominant product design, material, and processing methods – changed surprisingly little over 200 years. Steel tools on powered machines progressively removed metal chips to make metal parts.

The latter part of the 20th century saw a variety of fundamentally different metal-working methods based on new physical principles. These were potentially available for making firearms. When a mature and well understood technology is replaced by a less understood newer technology, how does the causal knowledge graph change, and what are the consequences? Since new physical technologies are critically important to long-term progress in most industries, we discuss several examples in more detail than warranted by their current importance to firearms manufacture.

New methods of precision machining – among them, electrical discharge, electrical chemical, abrasive water jet, and ultrasonic machining – employ entirely different physical principles to remove metal from workpieces.[7] Electrical discharge machining (EDM), for example, removes material by means of thermal energy generated by a spark across a gap between the tool and workpiece. The spark produces an extremely high temperature (up to 10,000°C) plasma channel

---

[7] Technology descriptions are from [31].

that evaporates a small amount of material. As the tool and workpiece move according to a computer-controlled trajectory, the spark shifts and the workpiece is shaped.

Removing material by vaporizing it with a spark clearly involves different variables than cutting it with a metal tool. EDM process performance is unaffected by the *hardness, toughness*, and *strength* of the material, but is affected by *melting temperature, thermal conductivity,* and *electrical conductivity,* the converse of the variables that matter in conventional machining. Dimensional tolerances of three microns can be achieved. Increasing the peak current can increase the machining rate, but the surface finish becomes rougher. Maximum production rates are also limited because at too high a power, tool wear becomes excessive, machining becomes unstable, and thermal damage occurs.

In electrical chemical machining (ECM), used to make rifle barrels among other applications, an electrolytic fluid is pumped between the tool and workpiece and current is applied to the tool. A variety of electrolytes can be used. The workpiece surface material dissolves into metal ions that are carried away by the fluid. ECM performance, like the performance of EDM, is unaffected by the *strength* and *hardness* of either tool or workpiece and is affected by their *electrical parameters*, but unlike EDM, performance is unaffected by their *thermal* behavior.

Why use ECM instead of EDM when both processes can machine any electrically conductive material? ECM can fabricate parts with low rigidity such as those with thin walls. It is much faster and gives better surface finish, but has poor accuracy because the pattern of electrical current flow with a given tool is influenced by many factors and difficult to predict. Tool shape must thus be modified by trial and error before making actual workpieces. Even then accuracy is only 10 to 300 microns, which is greatly inferior to that achieved with EDM.

Abrasive water jet machining involves spraying water mixed with abrasive particles onto a workpiece. The particles remove material. Typically, the water pressure and velocity are extremely high, approximately one million pounds per square inch and supersonic, respectively, so safety and noise issues are important. Abrasive water jet machining can shape ceramics and other nonconductive materials that ECM and EDM cannot, and which conventional machining has difficulty with. Not surprisingly, the characteristics, problems, and variables associated with abrasive water jet machining vary markedly from those encountered in ECM and EDM machining.

With such different variables and different physical principles, new processes start with less detailed causal knowledge graphs. Over time, new variables are identified as important, and new techniques developed to increase precision, speed, and other figures of merit. Table 3.3 shows some of the factors thought to influence ECM performance, very few of which are relevant to conventional machining. It is significant that trial and error are still required to choose the final ECM tool shape and, even then, the process is less accurate than other removal methods. This means that important portions of the causal network that determine final workpiece shape are not well understood. It was only recently, for example, that ECM accuracy was shown to improve with pulsed instead of continuous voltage. [31, p 13–32] Knowledge about ECM variables reported in Table 3.3 is at a much lower level than that about conventional machining, and overall the process remains much closer than conventional machining to art.

As always, knowledge develops by progressive exploration and refinement of the causal knowledge graph. Before a technology can be used, precision must be adequate and cost reasonable, at least for favorable applications. Once in use, a host of intermediate variables can be further improved as more is learned. For example, electrolysis is an undesirable side effect that corrodes the surface of parts in EDM (Figure 3.2).

Electrolysis, intrinsic to the early days of EDMing and continuing until the early 1990s, is caused by stray voltage from the cutting process interacting with contaminant in the dielectric fluid and attacking the workpiece. Electrolysis is particularly problematic when machining titanium, carbide, and stainless and mold steels, all of which suffer from poor surface integrity and shortened tool life due to the effects of electrolysis. They often require significant secondary machining operations and excessive polishing, which affect the overall accuracy of the machined part. Titanium turns

Subsystem	Key Determinants of ECM Performance
Electrical power system	**Current**
	Current areal density
	Voltage
	Pulse shape (on time, rise rate, etc.)
Electrolyte composition	**Aqueous or nonaqueous**
	Organic/inorganic; specific molecules
	Alkalinity
	Mixtures
	Contamination
	Passivating or nonpassivating
Electrolyte circulation system	**Flow rate**
	Pressure
	Temperature
	Concentration
Tool design; tool/workpiece geometry	Contour gradient
	Radii
	Flow path
	Flow cross section
	Tool feed rate

Table 3.3 Key variables affect electro-chemical machining; new variables in bold
(based on [31])

"blue," while stainless steel can be weakened by a thick recast
layer; tool steels rust; and carbide suffers degradation, the result
of cobalt binder depletion. [34]

Fig. 3.2 Defects caused by electrolysis of carbide [34]

The leading vendors of EDM machines developed power supplies that reduced electrolysis without reducing cutting speed. Other improvements included better filtration of the electrolyte fluid and a variety of power-saving methods.

Whereas the new material removal processes are entirely different from conventional machining and require extensive new knowledge, some of the other causal knowledge subnetworks in Figure 1.5 change only modestly. For example, dimensional measurement methods for finished parts are similar no matter how a part is produced. An ECM machine is still numerically controlled, and knowledge about how to control movement during cutting requires only moderate additions. The mathematics of feedback control can be adapted for ECM rather than redeveloped from scratch. In most cases, taking full advantage of the different characteristics of the new process involves changes to the ancillary subsystems, but the causal knowledge graph requires only moderate additional *knowledge*, even if the optimal *method* turns out to be quite different.

Finally, progress from art to science for new processes being developed today is markedly faster than was the case for conventional machining. First, there is less to learn. Second, the fundamentals of art-to-science transition established by Taylor, namely reductionism, using systems of quantitative equations to express knowledge, and learning by controlled experiments, are well known and much more refined than when Taylor used them.[8] Third, the firearms industry, having lagged in the adoption of ECM and other new processes, can take advantage of knowledge developed elsewhere. Indeed, much of the

---

[8] And of course other learning tools are also available that were not available at comparable points in the development of conventional machining, such as sophisticated statistical methods, automated monitoring and data collection, and process simulation.

relevant knowledge need not even be thoroughly understood within the industry, as tools based on it can be purchased from suppliers, a consequence of the modularity property of knowledge graphs.

4

From Art to Science

How should we characterize the evolution of manufacturing in the light of this examination of firearms over the course of two centuries? As Jaikumar showed, the central problem throughout the development of manufacturing has been achieving adequate process control. Once society moved beyond making unique items by hand, predictability, consistency, speed, and eventually versatility became key. All of these require control. Each successive epoch confronted problems whose solutions demanded new operating methods. Developing the solutions required deeper knowledge. The dual changes to procedures and their underlying knowledge constituted an evolution from art to science.

How procedures evolved is examined in Section 4.1. The regularity of changes in knowledge is considered in Section 4.2. The future of manufacturing is discussed in Section 4.3, as well as whether other activities of modern economies will reveal similar patterns.

4.1. Changes in Procedures

Procedures specify what actions to take and how to perform them. The companion paper described how production evolved from completely idiosyncratic activities before 1800 to an all but unmanned manufacturing plant in 1985 [15]. We can identify three major trends in procedures over these centuries: increasing specificity, increasing scope, and increasing depth in the causal network (Table 4.1).

To describe the first major trend in the evolution from art to science over the six epochs, how manufacturing activities have become more completely specified over time, we can construct a scale that measures the *formality* of procedures. At the extreme of zero procedure, no written or even mental plan of work exists; all actions are based on moment-by-moment decisions. At the opposite extreme of complete procedure, all activity is controlled by detailed written or programmed instructions. At intermediate points, high level instructions are specified, but details of implementation and responses to contingencies are left open. Over time, pictures became detailed written procedures. (Compare the sketches from the 18th century in Figure 2.3 of [15] with the 1950 operations sheet for the M1 rifle in Figure 6.3.) In parallel cams, jigs, and fixtures forced specified trajectories of machine motion – the principle of increasing mechanical constraint. The most formalized procedures can be realized in microprocessor-based systems, which require detailed instructions and allow for elaborate contingent behavior.

The second major trend was an ever expanding *scope* of activities governed by formal procedures. More machining subsystems (see Figure 1.5) were brought under explicit control. In the American System, for example, tools and methods were devised to control final inspection. In the Taylor epoch, proceduralization brought activities such as maintenance, tool making, and setup under formal control. With the debut of CIM and FMS, control of material flows, machine scheduling, and the translation of specifications from development into manufacturing were effected through computer programs. Untended operation of an FMS is possible today because virtually all normal activities including inspections, tool changes, and material movements are governed by programmed procedures.

The third trend we observe is increasing *depth* of control, measured by the number of generations controlled in the causal network. Consider an important intermediate variable such as cutting speed. Higher speed increases immediate output rates, but causes multiple problems. With greater knowledge, we can make a more sophisticated judgment of

Nature of change	Measure
Formality of control	Amount of detail used to specify procedures
Scope/extent of control	Breadth of control, such as number of subsystems actively controlled
Depth	Number of ancestral variables monitored or controlled for each key variable

Table 4.1 Evolution of procedures from art to science

proper cutting speed, eventually reaching real-time decision-making. Although measuring more variables is costly and controlling them even more costly, depth of control tends to increase over time for reasons we discuss later.

Full proceduralization of all activities has never been achieved and in a dynamic world would be disastrous. Even in high volume repetitive manufacturing rare, diverse, or extreme circumstances will occur, and will not be well understood. To attempt to fully proceduralize them is counterproductive. For example, the response to emergencies should be "shut down the machine and signal for assistance."

Moreover, the appropriate formality and scope of procedure fluctuates over time rather than increasing monotonically. When new processes or products are adopted, the initial level of knowledge is lower than before. Methods can be highly procedural and *bad* if knowledge is inadequate, as happened initially with NC machines. We now turn to the evolution of the underlying knowledge.

4.2. How Knowledge Evolved

Specific new knowledge was critical to each epoch, but changes in knowledge followed regular patterns from epoch to epoch. We can group the patterns, somewhat arbitrarily, into three categories. First, certain broad classes of problems recur, and make manufacturing inherently difficult. For example, more requirements are added over time. Second, there are classes of recurrent solutions, including the development of new mathematical methods for each epoch. Third, causal knowledge graphs themselves have structure, and the structure evolves in specific ways. We address each in turn.

4.2.1. Why is Manufacturing Hard? Sources of Problems

Presumably, every branch of human technology and endeavor has its own difficulties, but some are especially acute for manufacturing and came up in epoch after epoch.

Growing list of requirements. Additional outcome variables (system requirements) were added over time. Some were created by new product requirements propagating back through the causal network, such as the use of new raw materials.[1] In the modern era, emphasis increased on reducing side effects such as pollution, contamination, safety hazards, and noise.[2] Each new requirement forces rapid learning. Often, changes made to satisfy a new requirement interact with established portions of the process, leading to changes elsewhere.

Both tolerances and operating speeds had to improve simultaneously. Two fundamental manufacturing requirements are speed and precision/tolerance. At a given state of knowledge, operating speed can be traded off against conformance quality, including the tolerances achieved. A machine can be run slower to reduce its vibration; additional or more thorough inspection steps can be added to catch more problems; setups and calibrations can be done more often. Yet, both tighter tolerances and higher operating speeds were required in each epoch, for economic reasons. The only way to satisfy both was through better process control.

Control of more and smaller disturbances. Many important variables, such as the exact position of a tool relative to a cut surface, are influenced by dozens of ancestors. At a tolerance of 1/64th inch, many are too small to matter or even to detect, but at a few microns tolerance the number of relevant variables grows many-fold.

---

[1] Competitive dynamics drive many of these requirements; Beretta had to improve to keep up with other firms. This is the Red Queen paradox familiar to many industries – running faster and faster to stay in the same place.

[2] For example, in the 1980s Beretta engaged in a bitter fight to win a contract for a new US military sidearm. Winning required meeting a multitude of requirements, including local production.

Control of side effects. Wherever energy is applied in a process, it creates side effects such as heat, vibration, contamination, and electromagnetic interference that are transmitted through the local environment. Because of the sensitivity of high-precision operations, these can cause significant difficulties in disparate portions of the process. Control of heat during cutting, for example, is a side effect that has been a concern for more than a century, forcing more detailed understanding of its causes and effects (see [15], Figure 8.3). Taylor demonstrated the importance of coolant, but as tolerances tighten heat cannot be adequately removed from the cutting zone, so its effects must be compensated for. This requires much more knowledge than for cooling. And as operating speed increases, the magnitudes of side effects increase, even without considering tightening tolerances.

If we define side effects as "undesired descendants of a variable," there are also many direct side effects of process and machine designs. For example, a tool can be strengthened by making it larger, but this changes its thermal properties and requires larger motors to move it.

Solved problems may recur. A solution that is adequate at one level of performance may be inadequate when requirements such as speed of production change, or when side effects from elsewhere increase. When this happened, old solutions were refined and new ones added.

4.2.2. Measurement, Feedback, and Other Recurrent Solutions

Just as some problems are ubiquitous in manufacturing, some classes of solutions were vital to solving diverse problems across epochs.

New Mathematical Methods. New mathematical techniques supported the creation and articulation of knowledge in each epoch. Projective geometry ushered in the English System and simultaneous equations and custom slide rules were central to the calculations that were a hallmark of the Taylor system. Later epochs evolved on the back of statistical methods such as design of experiments in the SPC epoch, programming languages and Proportional-Integral-Derivative control in the NC epoch, and 3-D CAD and simulation techniques like FEM in the CIM epoch.

Strategies for Controlling Variation. Three generic strategies for controlling variation in a key variable are modifying the process to make it more robust, reducing variation in the most influential ancestors, and adding feedback control. An example is the problem of chatter discussed in Section 3, which can be solved by close operator monitoring of the machine, by reducing the speed of cutting, or by reinforcing the machine structure to reduce vibration.

Feedback-based control. Because causal knowledge graphs are never perfect renditions of the true causal network, all manufacturing depends on feedback, and increasing sophistication of feedback was a hallmark of evolution in procedures and knowledge. Consider an important intermediate variable $W = f(\mathbf{X}, \mathbf{V})$ where $\mathbf{X}$ and its effects on W are well understood but the constituents of $\mathbf{V}$, or the relationships $\partial f/\partial \mathbf{V}$, at low stages of knowledge. As $\mathbf{V}$ varies according to its own causal network, it creates stochastic variation in W with no visible cause. One solution is to learn more about f and $\mathbf{V}$, and learn to control the most important elements of $\mathbf{V}$, but this is time-consuming and expensive. The genius of feedback is that W can be partially stabilized without understanding $\mathbf{V}$, by manipulating one or more elements of $\mathbf{X}$ to compensate for measured changes in W. Feedback can also be used to reduce variation in W caused by parents in $\mathbf{X}$ that are known but expensive to control. Learning enough about $\mathbf{X}$ and $\partial f/\partial \mathbf{X}$ to use feedback thus constitutes a critical step in learning to control W. It can change W from adjustable (stage 3 of knowledge) to capable (stage 4). As a result, *feedback is a general technique for interrupting the propagation of variation downstream through a causal network*.

Improvements in measurement. Feedback has serious limitations, one of which arises from the fact that W cannot be measured perfectly. Control of a variable by feedback is bounded by how well that variable is measured, and the evolution of measurement knowledge has played a key role in the evolution of manufacturing. Even where direct feedback is not used, accurate measurement is needed for calibration, adjustment, verification, and especially for learning.

Measurement techniques are production processes for information, with their own causal knowledge graphs, so knowledge about metrology evolved according to the patterns described here. Although measurement technology often arrived from other industries, additional knowledge about how to use it effectively still had to be developed. For example, many measurement techniques are very sensitive to environmental disturbances, which are context dependent.

Accuracy, precision, repeatability, and related attributes are critical outcome variables for measurement processes. Less obvious is the importance of measurement *speed*. Because information turnaround time is a critical determinant of the effectiveness of feedback of all kinds, faster measurement enables better process control. If measurement takes several days, feedback cannot compensate for faster change such as diurnal or setup-caused. Faster measurement also increases the speed of subsequent learning [8, 38].

Measurement methods for a variable therefore tend to evolve through a sequence of techniques as metrology knowledge advances.[3]

- Measurement is generally first developed in a laboratory. To the extent that it is used in manufacturing, it is performed off-site using special equipment. This is acceptable because the variable is not measured routinely, but rather used in lab experiments.

- As metrology vendors develop special purpose tools embodying the new techniques, measurement is performed in specialized, on-site test labs. Although information turnaround can take days, such measurement can still be useful for field experiments, troubleshooting, checking incoming materials, and supporting various kinds of quality assurance, as well as calibrating production equipment and instruments.

- As technology progresses, measurement is performed on the factory floor in specific workstations. This was the norm for control charting and for measuring test pieces at the start of a batch.

- Measurement tools are built into machines, but the machine must be halted while a measurement is taken.

---

[3] Specifics of this sequence are based on unpublished notes on measurement in semiconductor and hard disk drive manufacturing.

- If a particular variable becomes important enough and is hard to control except by real-time feedback, measurement is ultimately performed while the machine is operating, with results available immediately.

Often, several different physical principles can be used to measure a variable. With different combinations of speed, precision, and cost, different methods are often employed concurrently at different locations. Economics plays a significant role in decisions about which variables to measure and how.

4.2.3. Structure and Evolution of Knowledge Graphs

Causal networks for actual working systems such as factories reflect the specifics of the design, construction, and operation of that system. But they are determined by natural laws, operating at levels from the atomic and nanosecond (chemical reactions and semiconductor gates) to tens of meters and days (inventory flows in a bulk processing plant). Knowledge graphs, which approximate the underlying causal network, are further constrained by the way people and organizations learn. Based on the cases discussed here and in [15] we can describe these graphs and how they changed.

Increasing local complexity. Knowledge graphs for individual phenomena become more complex over time. Added complexity includes first the addition of previously unrecognized variables, second ever deeper graphs comprising more generations of ancestors, and third a growing number of links due to discovering additional relationships among variables.

Rising stages of knowledge for variables and relationships. Discovering *new* variables and causal relationships changes the structure of the knowledge graph. But learning can also improve knowledge about previously identified variables and causal relationships. This does not change the structure of the graph, but does change the stage of knowledge of individual elements. Many variables that are eventually tightly controlled (i.e., at stage 4) were at one time only recognized (i.e., at stage 1). For example, Taylor's serendipitous discovery that tool steel could be improved by (what we now call) heat treatment took hardening from stage 0 (unknown) to stage 3 (adjustable). Decades of subsequent research led to a scientific model of the key relationships that brought knowledge to stage 4 (capable).

Multiple solutions. Because the causal networks that determine outcomes are complex, problems can usually be solved in multiple ways. For example, a causal path connecting a source of variation to a harmful effect can be interrupted at many links. Effectiveness, amount of new knowledge needed, and side effects vary for different solutions. New solutions are usually added, rather than replacing old ones.

Backwards evolution of knowledge graphs. Knowledge tends to evolve backwards. Deeper understanding of what causes a variable to vary hinges on a fuller understanding of parental relationships. Sometimes the parents can be partially controlled directly, but refining control of the parents requires understanding the grandparents, and so on.

Punctuated gradualism. Knowledge evolved by "punctuated gradualism," meaning incremental learning interspersed with occasional technological discontinuities. Incremental learning takes the form of gradual accretion of knowledge about phenomena, and gradual adjustments of procedures and tools. Discontinuities occur when the introduction of a new technique or requirement forces rapid learning about a host of new phenomena, and re-visiting many old variables, such as occured with electro-chemical machining.[4] Epochal shifts in manufacturing were marked by multiple discontinuities in parallel, but local discontinuities can occur at any time.

Causal networks are not tree-structured. It is convenient to model complex systems as hierarchical trees of systems and subsystems, and many authors including Vincenti have emphasized hierarchical decomposition of technological devices. However, causal networks are thickets and not trees.[5] That is, variables have multiple descendants and not just multiple parents. This makes them much more difficult to analyze and control. Changes to one variable, intended to produce a desired effect in a particular descendant, will also change many other

---

[4] This is a purely technological definition of disruptive change.

[5] Decompositions of systems into subsystems also differ from knowledge graphs in that links are not based on causal relationships.

descendants, often in undesirable ways. Environmental side effects are an example, but the phenomenon is much more general.

Modularity. As a partial substitute for tree-structure, manufacturing processes have some degree of modularity in their causal networks, and therefore in their knowledge graphs. This was demonstrated by Taylor's use of reductionism, which would not have worked without modularity (Section 2.3). In a near-modular network, although changing one variable will have many effects, most of the descendants are close to the original change. Closeness is measured by length of the causal path, but short paths usually correspond to physical closeness as well. In a modular process other subsystems can be ignored except for effects that propagate through the small number of relationships between modules.[6]

A very useful form of modularity is the sequential relationship among steps in a manufacturing process, such as raw material → shaped part → assembly → tested product. Each step can be treated as a module, with many internal causal links and few external links. Furthermore, causal paths that link different steps/modules can only occur through one of three mechanisms: information flows, environmental side effects, and by far the most important, physical transfer of work in process (WIP). So, if something goes wrong with an upstream process it can be detected, at least in principle, by looking at the properties of the WIP. Clever rearrangement of WIP can quickly isolate a problem to a single step/module.

Fractal nature of knowledge graphs. The more closely a causal system is examined, the more detail it contains. To a plant manager, a phenomenon such as rework might be summarized by a single variable, whereas a process engineer will have complex and evolving knowledge of the same phenomenon. On a very different scale, a process engineer can alter a machine's behavior by setting a few parameters in a PID controller, but the activity set in motion by those parameters includes electrons flowing through millions of gates inside a microprocessor. As a result, the patterns discussed above, such as punctuated gradualism, occur on multiple scales. To the plant accountant, rework

---

[6] [32] has a detailed discussion of hierarchy versus modularity in metabolic networks.

evolves smoothly, while to the process engineer it is the result of multiple discrete changes, some of them radical.

4.3. The Quest for Perfect Science

We have seen that the progression of manufacturing from art towards science consisted of advances in knowledge and the methods that embody it. Each epoch brought major improvements; rework at Beretta declined from more than fifty percent to less than one percent of activity. Can we predict that at some point knowledge will be perfect? To what extent can we say that manufacturing approaches the "end of its history," with complete understanding, absolute predictability, ideal performance, and nothing left to learn?

The answer is in two parts. Day-to-day production that exploits existing knowledge can approach this level. But dynamic tasks such as problem solving, design, and technology development, which extend knowledge, will always be a mixture of art and science.

For production in a static world, meaning a well-established process turning out a mature, thoroughly understood product for a known marketplace, it is feasible to bring processes to a level at which there is little left to learn and virtually all (normal) activity is highly procedural. The Taylor System was the apotheosis of this static view of manufacturing. Indeed, Taylor might be ecstatic about both how much is known by engineers today and how well procedures are executed by machines. Yet even when knowledge is virtually complete, some rare disruptions will necessitate human intervention.

But more fundamentally, the manufacturing world is not static. Competitive pressure, progress in upstream technologies such as materials science, and new features demanded by customers will inevitably drive the development of new products and new processes to produce them. Almost by definition, these products and processes will push the limits of what is known, and will therefore enter production only part way along the art-to-science spectrum.

Furthermore, the key tasks in a dynamic world are those of learning and problem solving, which are far from perfect science. Consider, for example, the problem of discovering and fixing a variety of intermittent problems that are detected at the end of a long production process. Questions that must be approached more as craft than procedure include: Which problems do we work on first? Who should be assigned to a particular problem? How should overall problem-solving efforts be organized? Having diagnosed a problem, where in its causal network should we attempt to fix it? How do we know we have actually solved a problem? When should we drop a problem and move on to something else? Are several different problems manifestations of single underlying problem?

Such questions involve ambiguity and uncertainty, and answering them requires expertise. Learning and problem solving, because they will continue to require human judgment and intuition, will never reach full procedure or full knowledge. Balconi in a study of recent changes in European manufacturing industries reaches the same conclusion:

> [T]raditional tacit skills of workers have become largely obsolete and modern operators on the shop floor are mainly process controllers and low-level problem solvers. Alongside this, the acceleration of innovation has made high-level problem solvers increasingly important. Tacit knowledge has thus remained crucial, but it has become complementary to a codified knowledge base and concerns problem solving heuristics, interpretation of data, etc ...

> In fact the performance and survival of firms depends on the individuals' ability to solve problems, to control, to improve processes, to find new technical solutions and to design new products, to integrate various "bodies of understanding" and to build relations with clients and interpret market trends. In conclusion, whereas the product of searching activity in the technological field is codified knowledge (know-how and know-what), it is the process of searching itself and of creating new artifacts which is embodied in individuals (depending on acts of insight). [6]

Yet, although learning activities will never reach the level of static manufacturing tasks, there was considerable progress toward science from Taylor to the present. Table 4.2 summarizes some of the important developments, most of which are soft tools that assist experts.

Currently, the leading industry for innovation in learning methods and tools is probably semiconductors. There, the economic rewards for faster development and ramp-up are high, and the high noise levels and long time lags in semiconductor fabrication mandate use of more procedural learning methods.[7]

Finally, the engineering disciplines of control theory and artificial intelligence have made modest progress towards formal computerized learning in well-structured systems. Adaptive control methods can compensate for minor design errors and component failures. Some of the more ambitious systems gradually moving from academia into the semiconductor industry use explicit causal models of the system being controlled. Both theoretical models (stage 4 of causal knowledge) and empirical fitting to statistical data (stage 3) are employed. Such systems can monitor a sequence of steps in photolithography, continuously monitoring the process and detecting out-of-control equipment. Although we can expect continued incremental progress towards automatic systems for refining coefficients when the structure of the causal network is already known, unstructured learning has resisted automation and is likely to do so for the foreseeable future.

4.3.1. Non-Manufacturing Applications

We have examined the evolution of manufacturing and the structure of the knowledge that supports it; but the structure of knowledge for some other technologies is similar. Consider the analogy between manufacturing and air transportation systems. A factory is a complex system organized to transform raw materials into useful products, quickly and precisely. An airline is a complex system involving aircraft, maintenance, airports, and air traffic control organized to move individuals from one location to another, quickly, precisely, and reliably. Both factory and airline are designed and operated using technological knowledge, specifically physical cause and effect relationships that can be modeled as causal knowledge graphs. Among the many analogies, both are heavily concerned with maintaining control despite variation

---

[7] This is a vast topic. [13; 10; 43]

Invention	Epoch of first use	Area of improvement*
Controlled experiments	Taylor	Signal to Noise ratio (S/N)
Systems analysis using reductionism	Taylor	Information Turnaround Time (ITAT)
Mathematical modeling of phenomena	Taylor	Cost, ITAT, S/N, generalizability beyond conditions tested
Statistical concepts and techniques (e.g. control chart, regression, experimental design)	SPC	S/N
Using science-based explanations of phenomena	SPC	Generalizability; use of outside knowledge
Computer simulation of processes or products	CIM/FMS	Cost, ITAT, S/N
Massive database of process variables (Factory Information System)	NC	Facilitates natural experiments e.g. data mining
Interaction between academic research and field problems	NC	Use of outside knowledge
Faster, more precise measurement methods	All	ITAT, S/N
* Principal impact of innovation on learning; terminology from [8]		

Table 4.2 Selected innovations in methods of learning

in the environment: weather for aviation, conditions inside the plant for manufacturing.

Given these similarities, it is not surprising that we see analogous patterns in the evolution of procedures and knowledge in the two sectors. Methods of flying, guiding, and maintaining aircraft have become more scientific with increasing scope, increasing depth, and increasing formality, exemplified by the "automated cockpit" of contemporary commercial aircraft. As far as knowledge about aviation, there are analogs of most of the evolutionary patterns discussed in Section 4.2, such as a growing list of requirements (passenger comfort; noise control at airports; anti-piracy), use of new mathematical methods (extensively discussed by Vincenti), feedback-based control, and improvements in measurement.

Vincenti has an excellent example of raising the knowledge stage of variables, specifically the struggle to identify variables that measure flying-qualities of different aircraft. This is defined as the ease and precision with which a pilot can control an aircraft. Initially, test pilots could express an opinion about an aircraft as "easy" or "hard" to control, but these subjective judgments were at stage 1 of knowledge. Decades of research were needed to fully define the key variables that should be used as formal specifications for designers. A report from 1937, for example, discusses how to measure 17 different variables during test flights of new aircraft. With hindsight, it is easy to overlook the initial confusion about the existence and definition of variables. Modern engineers routinely use a variable called "stick force per g" but "express amazement that any [other] maneuverability criterion ever existed and that it took [more than five years] to develop." [41, p 96]

Similarities in the evolution of aviation and manufacturing are not too surprising given the extreme dependence of both on physical processes. For a less similar industry, we might look, for example, at the back rooms of banks and other information processing "factories." Did such industries exhibit epochal shifts in the nature of work, from craft to functional specialization to statistical process control to, ultimately, process intelligence? Certainly we can point to many non-manufacturing industries in which managing intellectual assets is now critical, but historical research would be needed to investigate parallels with the intellectual shifts in firearms manufacture.

What about intellectual tasks such as design? The processes by which products are designed, the necessary supporting knowledge, and the tools employed by product designers all evidence evolution from art to science as we have defined it. As underlying knowledge about how products can be made to work becomes more elaborate, causal knowledge graphs grow. Design methods become more formal and procedural and portions of the design task more heavily automated. Draftsmen, for example, are no longer needed to translate design intent into engineering drawings and design calculations that used to employ slide rules now employ digital simulation.

The types of knowledge identified in Vincenti's studies of aeronautical design, however, go beyond the knowledge of physical cause and effect that we have considered. His engineers made extensive use of meta-knowledge – knowledge about how to manipulate causal knowledge to arrive at new designs (not his terminology). One of Vincenti's knowledge categories is *design instrumentalities*, the understanding of how to carry out the activities of design. One type of design instrumentality is structured design procedures such as optimization, satisficing, and deciding how to divide a system into subsystems. Less tangible design instrumentalities include ways of thinking such as visual thinking and reasoning by analogy, and judgmental skills such as intuition and imagination. [41, p 219ff]

Vincenti points out an equally important class of meta-knowledge, namely the methods used to extend causal knowledge, which he classifies into invention, transfer of knowledge from science, theoretical engineering research, design practice, experimental engineering research, production, and direct trial. The methods Taylor used to develop manufacturing knowledge correspond roughly to the last three methods in Vincenti's list. The first three were also relevant in various epochs, such as theoretical engineering research on the forces and geometry of metal cutting in the 1950s.

Learning methods and design instrumentalities are meta-knowledge about how to create and then exploit causal knowledge about underlying physical systems. As with physical manufacturing, specific design and learning tasks that used to require experts can now be done by soft tools. Nevertheless, because they will always depend partly on creativity and human intuition, design and especially learning will never approach perfect science.[8]

---

[8] A number of engineers, managers, and academics contributed to this research. Most important was R. Jaikumar, who under better circumstances would have co-authored this paper. He was my collaborator on many of its ideas. Special thanks to Jai's former editor, John Simon, for his work on the manuscript. Paul Dambre of Bekaert, who uniquely combines mastery of both scientific theory and manufacturing practice in his industry, patiently shared his expertise and was always an eager sounding board and experimenter. None of them bears responsibility for errors and omissions. REB

References

[1] H. G. J. Aitken, *Taylorism at Watertown Arsenal; scientific management in action, 1908–1915*, Harvard University Press, Cambridge, Mass., 1960.

[2] Maryam Alavi and Dorothy E. Leidner, "Review: Knowledge management and knowledge management systems: Conceptual foundations and research issues," *MIS Quarterly*, vol. 25, no. 1, pp. 107–, 2001.

[3] Anonymous, *The Theatre of the present war in the Netherlands and upon the Rhine: containing a description of all the divisions and subdivisions, rivers, fortified and other considerable towns, in the ten catholick provinces, the south-west part of Germany, the frontiers of France towards each, and all Lorrain, including the whole scene of military operations, that may be expected during the hostilities in those countries: with a general map, sixty eight plans of fortified places, and seventeen particular maps, upon a larger scale of the territories round most of the chief cities: also a short introduction to the art of fortification, containing draughts and explanations of the principal works in military architecture, and the machines and utensils necessary either in attacks or defences: also a military dictionary, more copious than has hitherto appear'd, explaining all the technical terms in the science of war*, J. Brindley, London, 1745.

[4] Diane E. Bailey and Julie Gainsburg, *Studying Modern Work: A "Knowledge Profile" of a Technical Occupation*, May 17, 2004.

[5] Margherita Balconi, *Codification of technological knowledge, firm boundaries, and "cognitive" barriers to entry*, DYNACOMP Working Paper, 2002.

[6] Margherita Balconi, "Tacitness, codification of technological knowledge and the organisation of industry," *Research Policy*, vol. 31, no. 3, pp. 357–379, 2002.

[7] Carl G. Barth, "Slide rules for the machine shop as a part of the Taylor system of management," *ASME*, vol. 25, pp. 49–62, 1904.

[8] Roger E. Bohn, *Learning by Experimentation in Manufacturing*, Harvard Business School, working paper 88-001, June, 1987.

[9] Roger E. Bohn, "Measuring and managing technological knowledge," *Sloan Management Review*, vol. 36, no. 1, pp. 61–73, 1994.

[10] Roger E. Bohn, "Noise and learning in semiconductor manufacturing," *Management Science*, vol. 41, no. 1, pp. 31–42, 1995.

[11] Roger E. Bohn and Ramchandran Jaikumar, "The development of intelligent systems for industrial use: An empirical investigation," In: *Research on Technological Innovation, Management and Policy*, Rosenbloom, Richard, JAI Press, Greenwich Connecticut, vol. 3, pp. 213–262, 1986.

[12] Edward W. Constant, "Why evolution is a theory about stability: Constraint, causation, and ecology in technological change," *Research Policy*, vol. 31, pp. 1241–1256, 2002.

[13] Daren Dance and Richard Jarvis, "Using yield models to accelerate learning curve progress," *IEEE Transactions on Semiconductor Manufacturing*, vol. 5, no. 1, pp. 41–46, 1992.

[14] Amy C. Edmondson, Ann B. Winslow, Richard M.J. Bohmer, and Gary P. Pisano, "Learning how and learning what: Effects of tacit and codified knowledge on performance improvement following technology adoption," *Decision Sciences*, vol. 32, no. 2, 2003.

[15] Ramchandran Jaikumar, "From filing and fitting to flexible manufacturing: A study in the evolution of process control," *Foundations and Trends® in Technology, Information and Operations Management*, vol. 1, no. 1, pp. 1–120, 2005.

[16] Ramchandran Jaikumar and Roger E. Bohn, "The development of intelligent systems for industrial use: A conceptual framework," In: *Research on Technological Innovation, Management, and Policy*, Rosenbloom, Richard S., JAI Press, London and Greenwich, Connecticut, vol. 3, pp. 169–211, 1986.

[17] Ramchandran Jaikumar and Roger E. Bohn, "A dynamic approach to operations management: An alternative to static optimization," *International Journal of Production Economics*, vol. 27, no. 3, pp. 265–282, 1992.

[18] Bruce Kogut and Udo Zander, "Knowledge of the firm and the evolutionary theory of the multinational corporation," *Journal of International Business Studies*, vol. 24, pp. 625–645, 1993.

[19] Ranga Komanduri, "Machining and grinding: A historical review of the classical papers," *Applied Mechanics Review*, vol. 46, no. 3, pp. 80–132, 1993.

[20] Thomas R. Kurfess, "Precision manufacturing," In: *The Mechanical Systems Design Handbook*, CRC Press, 2002.

[21] Robert G. Landers, A. Galip Ulsoy, and Richard J. Furness, "Process monitoring and control of machining operations," In: *The Mechanical Systems Design Handbook*, CRC Press, 2002.

[22] M. A. Lapré, A. S. Mukherjee, and L. N. Van Wassenhove, "Behind the learning curve: Linking learning activities to waste reduction," *Management Science*, vol. 46, no. 5, pp. 597–611, 2000.

[23] Michael A. Lapré and Luk N. Van Wassenhove, "Managing learning curves in factories by creating and transferring knowledge," *California Management Review*, vol. 46, no. 1, pp. 53–71, 2003.

[24] Steven Y. Liang, "Traditional machining," In: *Mechanical Engineering Handbook*, CRC Press LLC, Boca Raton, pp. 13-9–13-24, 1999.

[25] M. Eugene Merchant, "Mechanics of the metal cutting process I: Orthogonal cutting and the type 2 chip," *Journal of Applied Physics*, vol. 16, pp. 267–275, 1945.

[26] M. Eugene Merchant, "An interpretive review of 20th century US machining and grinding research," *Sadhana*, vol. 28, no. 5, pp. 867–874, 2003.

[27] Kazuhiro Mishina, "Learning by new experiences: Revisiting the flying fortress learning curve," In: *Learning by Doing in Markets, Firms, and Countries*, Lamoreaux, Naomi, Raff, Daniel, and Temin, Peter, University of Chicago Press, 1998.

[28] John Newton, *An introduction to the art of logick: Composed for the use of English schools, and all such who having no opportunity of being instructed in the Latine tongue, do however desire to be instructed in this liberal science*, Printed by E.T. and R.H. for Thomas Passenger at the Three Bibles on London Bridge and Ben. Hurlock over against St. Magnus Church, London, 1671.

[29] Judea Pearl, *Causality: Models, Reasoning, and Inference*, Cambridge University Press, Reprinted with corrections 2001, 2000.

[30] Michael Polanyi, *The Tacit Dimension*, Doubleday, 1967.

[31] K. P. Rajurkar and W. M. Wang, "Nontraditional machining," In: *Mechanical Engineering Handbook*, CRC Press LLC, Boca Raton, 1999.

[32] E. Ravasz, A. L. Somera, D. A. Mongru, Z. N. Oltvai, and A.-L. Barabasi, "Hierarchical organization of modularity in metabolic networks," *Science*, vol. 297, pp. 1551–1555, 2002.

[33] J. Sedger, *Art without Science, or, The Art of Surveying [Microform]: Unshackled with the Terms and Science of Mathematics, Designed for Farmers' boys*, Printed by Sampson, Chittenden & Croswell, Hudson, N.Y, 1802.

[34] Steve Szczesniak, "Anti-Electrolysis is pro EDM," *Modern Machine Shop*, 1998.

[35] Frederick W. Taylor, "On the art of cutting metals," *Transactions of the ASME*, vol. 28, pp. 31–248, 1907.

[36] Frederick W. Taylor, "The principles of scientific management," In: *Scientific Management (comprising Shop Management, The Principles of Scientific Management, and Testimony Before the Special House Committee)*, Harper & Brothers Publishers, New York, pp. xvi + 638, 1911.

[37] Christian Terwiesch and Roger E. Bohn, "Learning and process improvement during production ramp-up," *International Journal of Production Economics*, vol. 70, no. 1, pp. 1–19, 2001.

[38] Stefan H. Thomke, "Managing experimentation in the design of new products," *Management Science*, vol. 44, no. 6, pp. 743–762, 1998.

[39] E. M. Trent, *Metal Cutting*, Butterworths, London, 2nd edition, 1984.

[40] E. M. Trent and Paul Kenneth Wright, *Metal Cutting*, Butterworth-Heinemann, Boston, 4th edition, 2000.

[41] Walter G. Vincenti, *What Engineers Know and How They Know It*, Johns Hopkins University Press, Baltimore, 1990.

[42] Walter G. Vincenti, "Real-world variation-selection in the evolution of technological form: Historical examples," In: *Technological Innovation as an Evolu-*

tionary Process, Ziman, John, Cambridge University Press, Cambridge, pp. 174–189, 2000.

[43] Charles Weber and Eric von Hippel, *Knowledge Transfer and the Limits to Profitability: An Empirical Study of Problem-Solving Practices in the Semiconductor Industry*, PICMET, Portland, 2001.

[44] Robert S. Woodbury, *Studies in the History of Machine Tools*, MIT Press, 1972.